PROBABILITY

PROCEEDINGS OF SYMPOSIA
IN PURE MATHEMATICS
Volume XXXI

PROBABILITY

AMERICAN MATHEMATICAL SOCIETY
PROVIDENCE, RHODE ISLAND
1977

PROCEEDINGS OF THE SYMPOSIUM IN PURE MATHEMATICS
OF THE AMERICAN MATHEMATICAL SOCIETY

HELD AT THE UNIVERSITY OF ILLINOIS AT URBANA-CHAMPAIGN

URBANA, ILLINOIS
MARCH 1976

EDITED BY
JOSEPH L. DOOB

Prepared by the American Mathematical Society
with partial support from National Science Foundation contract no. 76-00261

Library of Congress Cataloging in Publication Data

CIP

Symposium in Pure Mathematics, University of Illinois
at Urbana-Champaign, 1976.
Probability.

(Proceedings of symposia in pure mathematics ; v. 31)
Bibliography: p.
1. Probabilities—Congresses. I. Doob, Joseph L.
II. American Mathematical Society. III. Title.
IV. Series.
QA273.A1S96 1976 519.2 77-2017
ISBN 0-8218-1431-1

AMS (MOS) subject classifications (1970). Primary 60G45, 60Jxx.

CONTENTS

Proceedings of Symposia in Pure Mathematics
Volume 31, 1977

SMALL RANDOM PERTURBATIONS OF DYNAMICAL SYSTEMS WITH REFLECTING BOUNDARY

R. F. ANDERSON AND STEVEN OREY

Sometimes information about second order elliptic differential operators can be obtained by studying the associated diffusion processes; these in turn can be investigated by means of Itô stochastic differential equations. Such an approach has been successful in studying situations in which the elliptic operator is a perturbation of a first order differential operator; the corresponding stochastic differential equation is then a perturbation of a dynamical system. The basic references are Ventcel′ and Freĭdlin **[5]** and Varadhan **[3]**, **[4]**.

Our purpose is to study such problems when the diffusion moves in a domain with boundary, and is subject to oblique reflection at the boundary. Such results are needed if one wants to consider boundary value problems with Neumann conditions.

Our results have appeared in full in **[1]**. The first part of **[1]** gives the basic theorems; the second part treats applications to differential equations. Our aim here is to give a brief exposition of the ideas of the first part.

Consider the stochastic differential equation

$$(1_\varepsilon) \qquad dY_t^{x,\varepsilon} = \varepsilon(\sigma_t \circ Y^{x,\varepsilon} dW_t + c_t \circ Y^{x,\varepsilon} dt) + b_t \circ Y^{x,\varepsilon} dt, \qquad Y_0^{x,\varepsilon} = x,$$

under the following assumptions. The process W is d-dimensional Wiener process, $x \in R^d$. The "coefficients" b_t, c_t, σ_t are functionals on the space $C_{[0,\infty)}(R^d)$ of continuous functions with range spaces, R^d, R^d, and $R^d \times R^d$ respectively; the coefficients are nonanticipating, i.e., b_t, c_t, σ_t are to be measurable with respect to the σ-field generated by the coordinate maps π_s, $0 \leqq s \leqq t$ (here $\pi_s(y) = y(s)$). Let $a_t = (\sigma_t \sigma_t^*)$. It is assumed that the coefficients are bounded and Lipschitz continuous (e.g., the ith component of b satisfies

AMS (MOS) subject classifications (1970). Primary 60H05, 60J60, 35B25, 35K20.

$$|b_t^i(\eta) - b_t^i(\xi)| \leqq k \max(\{|\eta_s - \xi_s| : 0 \leqq s \leqq t\}).$$

Denote by $\mathscr{F}_t$ the completion of the σ-field generated by the random variables W_s, $0 \leqq s \leqq t$. The *Markovian case* is that in which the coefficients are functions, $b_t(\eta) = b(\eta(t))$, $c_t(\eta) = c(\eta(t))$, $\sigma_t(\eta) = \sigma(\eta(t))$. Then the $Y^{x,\varepsilon}$ are Markov processes with differential generators

$$(*) \qquad \frac{1}{2}\varepsilon^2 \sum_{i,j=1}^{d} a_{ij}(x)\frac{\partial^2}{\partial x_i \partial x_j} + \varepsilon \sum_{i=1}^{d} c_i(x)\frac{\partial}{\partial x_i} + \sum_{i=1}^{d} b_i(x)\frac{\partial}{\partial x_i}.$$

If $q = (q_{ij})$ is a positive definite $d \times d$ matrix, $x \in R^d$, $y \in R^d$, define $(x, y)_q = \sum q_{ij}x_i x_j$, $\|x\| = (x, x)_q^{1/2}$. For $\varphi \in C_{[0,\infty)}(R^d)$ set

$$I_T(\varphi) = \int_0^T \|\dot\varphi_s - b_s \circ \varphi\|^2_{a_s^{-1}(\varphi)}\, ds$$

if φ is absolutely continuous, $I_T(\varphi) = \infty$ otherwise.

THEOREM A. (a) I_T: $C_{[0,T]}(R^d) \to R^1$ *is lower semicontinuous.*
(b) *For any open subset G of* $C_{[0,T]}(R^d)$

$$\varliminf_{\varepsilon \downarrow 0} 2\varepsilon^2 \log P[Y^{x,\varepsilon} \in G] \geqq -\inf\{I_T(\varphi) : \varphi \in G, \varphi_0 = x\}.$$

(c) *For any closed subset F of* $C_{[0,T]}(R^d)$

$$\varlimsup_{\varepsilon \downarrow 0} 2\varepsilon^2 \log P[Y^{x,\varepsilon} \in F] \leqq -\inf\{I_T(\varphi) : \varphi \in F, \varphi_0 = x\}.$$

REMARKS. (1) This theorem is given in **[5]** for the Markovian case. The proofs go over without change to the more general situation.

(2) The specific form of the theorem given here is due to **[4]**, where the Markovian case with $b \equiv c \equiv 0$ was treated.

(3) Somewhat more information is given in **[1]**, with some of the refinements originally due to Friedman **[2]**.

Consider now a diffusion $X^{x,\varepsilon}$ starting at x, which moves in the interior of a region $D \subseteq R^d$ according to the differential generator $(*)$ and on reaching the smooth boundary ∂D of D is reflected instantaneously in the direction γ, where γ is a smooth vector field on ∂D directed strictly into the interior of the region. According to S. Watanabe **[6]**, such a process $X_t^{x,\varepsilon}$ together with its "local time on the boundary" $\xi_t^{x,\varepsilon}$ can be characterized as continuous processes, adapted to $(\mathscr{F}_t)$ and satisfying

$$(2_\varepsilon) \qquad \begin{aligned} dX_t^{x,\varepsilon} &= \varepsilon(\sigma(X_t^{x,\varepsilon})dW_t + c(X_t^{x,\varepsilon})dt) + b(X_t^{x,\varepsilon})dt + \chi_{\partial D}(X_t^{x,\varepsilon})\,\gamma(X_t^{x,\varepsilon})d\xi_t^{x,\varepsilon}, \\ X_0^{x,\varepsilon} &= x, \quad \xi_0^{x,\varepsilon} = 0 \end{aligned}$$

and $\xi_t^{x,\varepsilon}$ is nondecreasing in t, increasing only on $\Delta \equiv \{t : X_t^{x,\varepsilon} \in \partial D\}$, where Δ must have Lebesgue measure zero. We construct such a pair $(X_t^{x,\varepsilon}, \xi_t^{x,\varepsilon})$ by treating first the case where D is the half-space R_+^d and γ is the unit inward normal; the general case is then obtained by localisation.

We describe the construction in the half-space situation, since this is fundamental to our approach. (For Brownian motion, i.e., $b \equiv c \equiv 0$, a the identity, this reduces to a classical method of P. Lévy.) Define a transformation Γ of $C_{[0,\infty)}(R^d)$ into $C_{[0,\infty)}(R_+^d)$ by $\Gamma\zeta = \eta$, where the ith components are given by $\eta^i = \zeta^i$, $i =$

2, 3, $\cdots$, d, and $\eta_t^1 = \zeta_t^1 - \inf_{0\leqq s\leqq t}(\zeta_s^1 \wedge 0)$, and define $\Gamma_t(\zeta) = [\Gamma(\zeta)]_t$, $\Gamma(\zeta) - \zeta = (\xi(\zeta), 0, \cdots, 0)$, $\xi_t(\zeta) = [\xi(\zeta)]_t$.

PROPOSITION. *Let $Y^{x,\varepsilon}$ be the solution of (1_ε), where $x \in R_t^d$, $\varepsilon > 0$. Let $\sigma_t = \sigma \circ \Gamma_t$, $c_t = c \circ \Gamma_t$, $b_t = b \circ \Gamma_t$. Then $X^{x,\varepsilon} = \Gamma \circ Y^{x,\varepsilon}$, $\xi^{x,\varepsilon} = \xi \circ Y^{x,\varepsilon}$ satisfies (2_ε) and the associated conditions. Any other pair of processes satisfying (2_ε) and the associated conditions must agree with $(X^{x,\varepsilon}, \xi^{x,\varepsilon})$ with probability one.*

Return now to the "general" region $D \subseteq R^d$ with vector field γ. For $\varphi \in C_{[0,T)}(D)$ (continuous functions with values in D), set

$$I_T^+(\varphi) = \inf\{I_T(\psi): \psi \in C_{[0,T]}(R^d), \Gamma(\psi) = \varphi\}$$
$$= \int_0^T \left\|\dot\varphi_t - b(\varphi_t) - X_{\partial D}(\varphi_t)\cdot[\mathrm{comp}_\gamma^{a^{-1}}(\dot\varphi_t - b(\varphi_t))]^- \cdot \gamma(\varphi_t)\right\|a^{-1}(\varphi_t)\,dt.$$

In this formula $\mathrm{comp}_\gamma^{a^{-1}}(\cdot)$ means the component in the sense of the inner product $(x, y)_{a^{-1}}$ in the direction $\gamma(\varphi_t)$; and $\lambda^- = (\lambda \wedge 0)$.

THEOREM 1. *The results of Theorem A carry over to the reflected diffusion $X^{x,\varepsilon}$, if I_T^+ is used in place of I_T and D and $C_{[0,T]}(D)$ take the place of R^d and $C_{[0,T]}(R^d)$ respectively.*

REMARK. (4) For the half-space case this follows easily from Theorem A by the construction of $X^{x,\varepsilon}$. The general case requires localisation arguments.

Consider now an equation (2_0) obtained from (2_ε) by setting $\varepsilon = 0$, with associated conditions as before, *except* that it is no longer required that Δ have Lebesgue measure zero. Also introduce

$$\text{(3)}\qquad \begin{aligned} \xi_t^{x,0} &= -\int_0^t X_{\partial D}(X_s^{x,0})[\mathrm{comp}_\gamma^D\, b(X_s^{x,0})]^-\,ds,\\ X_t^{x,0} &= x + \int_0^t b(X_s^{x,0}) - X_{\partial D}(X_s^{x,0})[\mathrm{comp}_\gamma^D\, b(X_s^{x,0})]^-\,\gamma(X_s^{x,0})\,ds,\end{aligned}$$

where $\mathrm{comp}_\gamma^D(\cdot)$ means the component of the vector in the direction $\gamma(X_s^{x,0})$ when this vector is analyzed into a component along $\gamma(X_s^{x,0})$ plus a component tangential to ∂D.

THEOREM 2. *There exists a unique pair $(X^{x,0}, \xi^{x,0})$ solving (2_0) and the associated conditions. This pair satisfies (3). Also $X^{x,0}$ is the unique φ such that $\varphi_0 = x$, $I_T^+(\varphi) = 0$, $0 < T < \infty$.*

COROLLARY. (a) *For $\delta > 0$, there exists $\alpha > 0$ such that*

$$\overline{\lim_{\varepsilon\downarrow 0}}\, 2\varepsilon^2 \log P\left[\sup_{0\leqq s\leqq T}\left|X_s^{x,\varepsilon} - X_s^{x,0}\right| > \delta\right] < -\alpha.$$

(b) $\sup_{0\leqq s\leqq T}\left|X_s^{x,\varepsilon} - X_s^{x,0}\right| \to 0$ *in $L_p(P)$, $1 \leqq p < \infty$, uniformly for $x \in D$.*
(c) $\sup_{0\leqq s\leqq T}\left|\xi_s^{x,\varepsilon} - \xi_s^{x,0}\right| \to 0$ *in $L_p(P)$, $1 \leqq p < \infty$, uniformly for $x \in D$.*

REFERENCES

1. R. F. Anderson and S. Orey, *Small random perturbations of a dynamical system with reflecting boundary*, Nagoya Math. J. **60** (1976), 189–216.
2. A. Friedman, *Small random perturbations of dynamical systems and applications to parabolic*

equations, Indiana Univ. Math. J. **24** (1974/75), 533–553; erratum, ibid. **24** (1974/75), 903. MR **51** #4432.

3. S. R. S. Varadhan, *On the behavior of the fundamental solution of the heat equation with variable coefficients,* Comm. Pure Appl. Math. **20** (1967), 431–455. MR **34** #8001.

4. ———, *Diffusion processes in a small time interval,* Comm. Pure Appl. Math. **20** (1967), 659–685. MR **36** #970.

5. A. D. Ventcel′ and M. I. Freĭdlin, *On small random perturbations of dynamical systems,* Uspehi Mat. Nauk **25** (**151**) (1970), no. 1, 3–55 = Russian Math. Surveys **25** (1970), no. 1, 1–55. MR **42** #2123.

6. S. Watanabe, *On stochastic differential equations for multi-dimensional diffusion processes with boundary conditions.* I, II, J. Math. Kyoto Univ. **11** (1971), 169–180; ibid. **11** (1971), 545–551. MR **43** #1291; **44** #4815.

University of Pittsburgh

University of Minnesota

Proceedings of Symposia in Pure Mathematics
Volume 31, 1977

BROWNIAN MOTION AND CLASSICAL ANALYSIS

D. L. BURKHOLDER*

Some of the most exciting applications of probability theory during the last few decades have been to potential theory, differential equations, harmonic analysis, the structure theory of Banach spaces, and to other diverse areas of mathematics, some of which at one time seemed completely unrelated to probability.

Our aim here is not to survey such recent developments but rather to illustrate in a specific problem area some of the interplay between classical analytic concepts and probability theory.

Let R be an open, connected subset of $\boldsymbol{R}^n$ $(n \geqq 2)$, X a Brownian motion in $\boldsymbol{R}^n$ starting at a point x in R, and τ the first time X leaves R:

$$\tau(\omega) = \inf\{t > 0: X_t(\omega) \notin R\}.$$

What is the distribution of τ? If τ is finite, what is the distribution of X_τ, the place of the first exit? For most regions R, these questions are unrealistic. In fact, even for quite reasonable regions, the much less demanding question of whether or not the distribution of X_τ (harmonic measure) is absolutely continuous with respect to surface measure on the boundary ∂R of R is extremely difficult. For example, only recently has anyone been able to show that, for Lipschitz regions, harmonic measure and surface measure are mutually absolutely continuous (Dahlberg **[6]**).

Fortunately, in many applications, less than full knowledge of the distribution of τ (or of X_τ) is needed. Here we consider the pth moments of $\tau^{1/2}$ for $0 < p < \infty$, and show how they can be fruitfully related to some concepts and problems of classical analysis. Most of the results described here appeared first in **[4]**, where proofs and further details are given. The results of §4 are new.

AMS (*MOS*) *subject classifications* (1970). Primary 60J65, 31B05, 30A78; Secondary 60G40, 31A35.

Key words and phrases. Brownian motion, first exit time, harmonic majorization, boundary behavior, symmetrization, conformal mapping, Hardy spaces.

*Work supported by NSF.

1. Exit times and harmonic majorization. Throughout the paper, p denotes a positive real number and $E_x\tau^{p/2}$ the expectation of $\tau^{p/2}$ as a function of the starting point x of the Brownian motion X. We shall write $\tau^{1/2} \in L^p$ if $E_x\tau^{p/2}$ is finite for some $x \in R$ and, hence, for all $x \in R$.

THEOREM 1.1. *The exit time τ of the region R satisfies $\tau^{1/2} \in L^p$ if and only if there is a function u harmonic in R such that $|x|^p \leqq u(x)$, $x \in R$.*

So here we see that a difficult probability problem, whether or not $\tau^{1/2} \in L^p$, is equivalent to an often much simpler problem of classical analysis, whether or not $|x|^p$ is majorized in R by a ha monic function.

The mapping $x \to |x|^p$ is subharmonic in $\boldsymbol{R}^n$. Therefore, the existence of a harmonic majorant in R implies the existence of a least harmonic majorant. This leads beyond the qualitative result of Theorem 1.1 to

THEOREM 1.2. *If u is the least harmonic majorant of $|x|^p$ in R, then*

$$c_{p,n}E_x(n\tau + |x|^2)^{p/2} \leqq u(x) \leqq C_{p,n}E_x(n\tau + |x|^2)^{p/2}, \qquad x \in R.$$

As indicated, the choice of the constants in this two-sided inequality depends only on p and n. For the optimal choice,

$$\lim_{n\to\infty} c_{p,n} = \lim_{n\to\infty} C_{p,n} = 1 \tag{1.1}$$

and the convergence is uniform for p in bounded subintervals of $(0, \infty)$. Furthermore, $c_{2,n} = C_{2,n} = 1$ and

$$\sup_{0<p<\infty} \frac{C_{p,n} + c_{p,n}^{-1}}{p^{p/2} + 1} < \infty. \tag{1.2}$$

If $|x|^p$ does have a harmonic majorant in R or, equivalently, if $\tau^{1/2} \in L^p$, then the least harmonic majorant of $|x|^p$ in R is precisely $u(x) = E_x|X_\tau|^p$. In any case, a two-sided inequality exists,

$$E_x|X_\tau|^p \approx E_x(n\tau + |x|^2)^{p/2},$$

provided either $n \geqq 3$ and $P_x(\tau < \infty) = 1$, or $n = 2$ and $E_x \log \tau < \infty$. So finding the moments of X_τ is nearly the same problem as finding the moments of $\tau^{1/2}$.

The following sufficient condition for a function u to be the least harmonic majorant is sometimes easy to apply.

THEOREM 1.3. *If u is continuous on the closure of R with $u(x) = |x|^p$, $x \in \partial R$, and is harmonic in R with*

$$c|x|^p - c \leqq u(x) \leqq C|x|^p + C, \qquad x \in R,$$

then u is the least harmonic majorant of $|x|^p$ in R.

Here, as elsewhere in this paper, c and C denote positive real numbers.

We now illustrate these theorems with a simple application. If $0 < \alpha \leqq \pi$, let

$$R_\alpha = \{x \in \boldsymbol{R}^n : x \neq 0, 0 \leqq \theta < \alpha\}$$

where θ is the angle between the vectors x and $(1, 0, \cdots, 0)$. Let τ_α be the corresponding exit time.

First, consider the simplest case $n = 2$ and let $p\alpha < \pi/2$. Then, relative to R_α,

$$u(x) = |x|^p \cos p\theta / \cos p\alpha$$

satisfies the conditions of Theorem 1.3. Therefore, u is the least harmonic majorant of $|x|^p$ in R_α so, by Theorem 1.2,

$$E_{(1,0)}(2\tau_\alpha + 1)^{p/2} \approx 1/\cos p\alpha.$$

Accordingly,

$$\tau_\alpha^{1/2} \in L^p \Leftrightarrow p\alpha < \pi/2. \tag{1.3}$$

Now let $n \geqq 3$. Here the mapping $\theta \to \cos p\theta$ is replaced by

$$h(\theta) = F(-p, p + n - 2; (n - 1)/2; (1 - \cos\theta)/2)$$

where $F(a, b; c; t)$ is the hypergeometric function. Let $\theta_{p,n}$ denote the smallest positive zero of h in $(0, \pi)$. Then, for $\alpha < \theta_{p,n}$,

$$u(x) = |x|^p h(\theta)/h(\alpha)$$

satisfies the conditions of Theorem 1.3 so

$$E_{(1,0,\cdots,0)}(n\tau_\alpha + 1)^{p/2} \approx 1/h(\alpha).$$

Therefore,

$$\tau_\alpha^{1/2} \in L^p \Leftrightarrow \alpha < \theta_{p,n}.$$

See **[4]** for further details and the proofs of Theorems 1.1, 1.2, and 1.3. Note that although the statement of Theorem 1.3 contains no concepts from probability theory, the proof is probabilistic. The proofs rest on the basic work of Doob **[7]** and the inequality, obtained in **[5]**, between the maximal function and the quadratic variation of a Brownian motion up to a stopping time. Here we need the inequality in the following form: If τ is any stopping time of the Brownian motion X and $X_\tau^* = \sup_{0<t<\infty} |X_{\tau\wedge t}|$, then

$$c_{p,n}E_x(n\tau + |x|^2)^{p/2} \leqq E_x(X_\tau^*)^p \leqq C_{p,n}E_x(n\tau + |x|^2)^{p/2}$$

and the constants satisfy (1.1) and (1.2); see **[4]** and **[3]**. In particular,

$$\tau^{1/2} \in L^p \Leftrightarrow X_\tau^* \in L^p. \tag{1.4}$$

Expectations $E_x\Phi(\tau^{1/2})$ more general than $E_x\tau^{p/2}$ can also be studied. Here is one illustrative result:

THEOREM 1.4. *Let Φ be a continuous, nondecreasing function from* $[0, \infty]$ *onto* $[0, \infty]$ *satisfying the growth condition $\Phi(2\lambda) \leqq c\Phi(\lambda)$, $\lambda > 0$. Suppose u is continuous on $R \cup \partial R$, harmonic in R,*

$$\sup_{x\in\partial R} u(x) < \sup_{x\in R} u(x) \quad \textit{and} \quad |u(x)| \leqq C\Phi(|x|) + C, \qquad x \in R.$$

Then $\Phi(\tau^{1/2}) \notin L^1$.

For the method of proof and related results, see **[4]**.

If $\Phi(\lambda) = \log(1 + \lambda)$, $R = \{x \in \boldsymbol{R}^2\colon |x| > \delta\}$ for some $\delta > 0$, and $u(x) =$

$\log|x|$, the theorem gives $\Phi(\tau^{1/2}) \notin L^1$. This implies $\log^+\tau \notin L^1$, a well-known result. Or if $\Phi(\lambda) = \lambda^p$, R is the region R_α of (1.3) with $p\alpha = \pi/2$, and $u(x) = |x|^p \cos p\theta$, then $\tau_\alpha^{1/2} \notin L^p$.

For a third example, let $\Phi(\lambda) = \lambda^{1/2}$,

$$R = \{x = (x_1, x_2) \in \boldsymbol{R}^2: |x| > 2\delta - x_1\} \tag{1.5}$$

for some $\delta > 0$, and $u(x) = |x|^{1/2} \cos \theta/2$. Note that R is the region to the right of the parabola $x_2^2 = 4\delta(\delta - x_1)$. Here $u(x) = \delta^{1/2}$, $x \in \partial R$, so $\tau^{1/2} \notin L^{1/2}$.

2. Embedding. We now replace our original exit-time problem with an equivalent problem in higher dimensions. Let $R \subset \boldsymbol{R}^n$ and τ be as above. Suppose that N is a positive integer greater than n and consider the region $R_N = R \times \boldsymbol{R}^{N-n}$ of $\boldsymbol{R}^N$ and its exit time τ_N. Clearly, τ_N has the same distribution as τ. Suppose that $\tau^{1/2} \in L^p$. Then $E_{(x,0)}\, \tau_N^{p/2} = E_x \tau^{p/2}$ is finite so there is a least harmonic majorant u_N of $|\cdot|^p$ in R_N. We can now take advantage of the fact that the constants in the inequality of Theorem 1.2 approach unity as the dimension of the space increases.

THEOREM 2.1. *If* $\tau^{1/2} \in L^p$, *then* $E_x\tau^{p/2} = \lim_{N\to\infty} N^{-p/2} u_N(x, 0)$, $x \in R$.

PROOF. By Theorem 1.2,

$$N^{-p/2} u_N(x, 0) \approx E_{(x,0)}(\tau_N + |x|^2/N)^{p/2} = E_x(\tau + |x|^2/N)^{p/2} \to E_x\tau^{p/2}.$$

So the desired result follows from (1.1).

3. Symmetrization. Suppose the region $R \subset \boldsymbol{R}^n$ contains the origin and R_s is the region obtained from R by spherical symmetrization. That is, $R_s \subset \boldsymbol{R}^n$ contains the origin and has the following properties: If $r > 0$ and $\{|x| = r\} \subset R$ then $\{|x| = r\} \subset R_s$. If $r > 0$ and $\{|x| = r\} \not\subset R$, then

$$R_s \cap \{|x| = r\} = \{x \in \boldsymbol{R}^n: |x| = r, \theta < \alpha\}$$

where θ is the angle between the vectors x and $(1, 0, \cdots, 0)$ and α is chosen so that $R_s \cap \{|x| = r\}$ and $R \cap \{|x| = r\}$ have the same spherical Lebesgue measure. Let τ_s be the exit time of R_s. Then

$$P_0(X_\tau^* > \lambda) \leqq P_0(X_{\tau_s}^* > \lambda), \quad \lambda > 0,$$

which is a translation into the language of Brownian motion of an inequality for harmonic measure obtained by Baernstein [1] in the case $n = 2$, and by Baernstein and Taylor [2] in higher dimensions. Therefore,

$$X_{\tau_s}^* \in L^p \Rightarrow X_\tau^* \in L^p,$$

which implies, by (1.4), that

$$\tau_s^{1/2} \in L^p \Rightarrow \tau^{1/2} \in L^p.$$

If $R_s \subset R_\alpha$, where R_α is defined in §1, then

$$\tau_\alpha^{1/2} \in L^p \Rightarrow \tau_s^{1/2} \in L^p \Rightarrow \tau^{1/2} \in L^p, \tag{3.1}$$

a fact that will be useful in the next section.

4. Boundary behavior. If R is a bounded region of $\boldsymbol{R}^n$ and x_0 is a regular boundary point of R, then

$$\lim_{x \to x_0;\, x \in R} E_x \tau = 0.$$

This is well known; see, for example, Dynkin and Yushkevich [**8**]. Recently, Frank Knight asked us if this is also true for any region R such that $E_x \tau$ is finite for $x \in R$.

Here we answer this question in the negative by constructing a region

$$R \subset \{x = (x_1, \cdots, x_n) \in \boldsymbol{R}^n : x_1 > 0\} \tag{4.1}$$

with $\tau^{1/2} \in L^2$ such that every point x_0 on the hyperplane $\{x: x_1 = 0\}$ is a regular boundary point of R satisfying

$$\limsup_{x \to x_0;\, x \in R} E_x \tau = \infty. \tag{4.2}$$

This, incidentally, shows that a solution of the Poisson equation $\Delta v = -2$ in R can have the same bad behavior at the boundary: By Theorem 1.2, in which $c_{p,n} = C_{p,n} = 1$ if $p = 2$, the function $u(x) = E_x(n\tau + |x|^2)$ is harmonic in R so that $v(x) = E_x \tau = (u(x) - |x|^2)/n$ satisfies $\Delta v = -2$ there.

Now consider the boundary behavior of $E_x \tau^{p/2}$. The following two results, one negative, the other positive, illustrate some of the possibilities.

THEOREM 4.1. *There is a region $R \subset \boldsymbol{R}^n$, with all moments of its exit time τ finite, such that every point x_0 on the nonpositive x_1-axis,*

$$\{x = (x_1, \cdots, x_n) \in \boldsymbol{R}^n : x_1 \leqq 0, x_2 = 0, \cdots, x_n = 0\},$$

is a regular boundary point of R satisfying

$$\limsup_{x \to x_0;\, x \in R} E_x \tau^{p/2} = \infty \tag{4.3}$$

for all $p > 0$ if $n \geqq 3$, and for all $p \geqq 1/2$ if $n = 2$.

THEOREM 4.2. *Let $0 < p < 1/2$. If R is a region of $\boldsymbol{R}^2$ such that $\tau^{1/2} \in L^p$, then, for every regular boundary point x_0 of R,*

$$\lim_{x \to x_0;\, x \in R} E_x \tau^{p/2} = 0.$$

Before proving these two theorems, we shall construct a region R satisfying (4.2). The proof of Theorem 4.1 depends on a similar but somewhat more complicated construction.

EXAMPLE 4.1. Let $\cdots < \alpha_2 < \beta_2 < \alpha_1 < \beta_1 < \delta_1 < \delta_2 < \cdots$ be positive real numbers such that $\beta_k \to 0$ and $\delta_k \to \infty$, and let $a_k = (\alpha_k, 0, \cdots, 0)$, $b_k = (\beta_k, 0, \cdots, 0)$, and $d_k = (\delta_k, 0, \cdots, 0)$, be the corresponding points of $\boldsymbol{R}^n$. Let $B_1 \subset B_2 \subset \cdots$ be the (open) balls with centers on the positive x_1-axis such that the hyperplane $\{x: x_1 = \alpha_k\}$ is tangent to B_k at a_k and $\{x: x_1 = \delta_k\}$ is tangent to B_k at d_k. Let $R_1 = B_1$ and $R_{k+1} = B_{k+1} - (B_k \cup \partial B_k)$, $k \geqq 1$. Note that b_k belongs to R_k. Let $A_k = \{x \in \partial B_k : |x - d_k| < \varepsilon_k\}$ where $0 < \varepsilon_k < \delta_k - \alpha_k$. Finally, let $R = R_1 \cup A_1 \cup R_2 \cup A_2 \cup \cdots$. This is a region satisfying (4.1) such that every point on the hyperplane $\{x: x_1 = 0\}$ is a regular boundary point of R. We now show the δ_k can be chosen large enough and the ε_k small enough so that (4.2) holds with $\tau^{1/2} \in L^2$.

Let τ_k be the exit time of R_k. First, δ_1 can be chosen so large that $E_{b_1}\tau_1 > 1$. To see this, note that $R_1 \to \{x: x_1 > \alpha_1\}$ as $\delta_1 \to \infty$ and the exit time of a half-space has infinite expectation (for example, set $\alpha = \pi/2$ in (1.3) and use embedding). Next, δ_2 can be chosen so that $E_{b_2}\tau_2 > 2$ since, as $\delta_2 \to \infty$, R_2 converges to a region containing the half-space $\{x: x_1 > \delta_1\}$; hence the limiting region must also have an exit time with infinite expectation. So the δ_k can be chosen inductively to satisfy $E_{b_k}\tau_k > k$. Since $\tau \geqq \tau_k$, which follows from $R \supset R_k$, this implies that

$$E_{b_k}\tau > k, \qquad k \geqq 1. \tag{4.4}$$

Now choose ε_k so small that

$$P_{b_1}(X \text{ hits } A_k \text{ before } \partial B_k - A_k) \leqq e^{-\delta_{k+1}}.$$

This is possible. If $\delta_k \leqq \lambda < \delta_{k+1}$, the above event is implied by $X_\tau^* > \lambda$, so

$$P_{b_1}(X_\tau^* > \lambda) \leqq e^{-\delta_{k+1}} \leqq e^{-\lambda}.$$

Therefore, $X_\tau^* \in L^p$ and, by (1.4), $\tau^{1/2} \in L^p$ for all $0 < p < \infty$.

Now let x_0 be a point on the hyperplane $\{x: x_1 = 0\}$. If $x_0 = 0$, then (4.2) follows from (4.4). So suppose $r = |x_0| > 0$. We shall show there is a sequence of points $y_k \in R$ such that $|y_k| \to r$ and

$$E_{y_k}\tau > k/2 \tag{4.5}$$

for all large k. The region R is symmetric around the x_1-axis so by rotating the point y_k around this axis we can obtain a new point x_k in R satisfying both $E_{x_k}\tau = E_{y_k}\tau$ and $x_k \to x_0$. This implies (4.2).

Let σ be the exit time of the ball B with center 0 and radius r. Let k be large enough so that $b_k \in B \cap R_k$ and $k/2 < k - r^2/n$. Finally, let $v(x) = E_x\tau$ and I denote the indicator function of the set $\{\tau > \sigma\}$. Then by (4.4), the strong Markov property, and Theorem 1.2,

$$k < E_{b_k}\tau \leqq E_{b_k}[(\tau - \sigma)I] + E_{b_k}\sigma = E_{b_k}[v(X_\sigma)I] + (r^2 - |b_k|^2)/n.$$

So at some point in the probability space, $k/2 < k - r^2/n < v(X_\sigma)I = v(X_\sigma)$. Let y_k be the value of X_σ at this point. Then $y_k \in (\partial B) \cap R_k$ so $|y_k| \to r$ and (4.5) is proved.

PROOF OF THEOREM 4.1. Let α_k, β_k, δ_k and a_k, b_k, d_k be as at the beginning of Example 4.1. Here let B_k be the set of all $x \in \boldsymbol{R}^n$ such that $|x| < \delta_k$ and the distance from x to the nonpositive x_1-axis is greater than α_k. Define the R_k and the A_k in terms of the B_k as in Example 4.1. Let

$$R = R_1 \cup A_1 \cup R_2 \cup A_2 \cup \cdots.$$

This is the desired region provided the δ_k are chosen sufficiently large and the ε_k are chosen sufficiently small.

If $n = 2$, let $p_k = 1/2$; if $n \geqq 3$, let $p_k \to 0$. Let τ_k be the exit time of R_k. Then the δ_k can be chosen so that $E_{b_k}\tau_k^q > k$ where $q = p_k/2$. This is possible because, as $\delta_k \to \infty$, the region R_k converges to a region containing $\{x \in \boldsymbol{R}^n$: the distance from x to the nonpositive x_1-axis is greater than $\delta_{k-1}\}$ and the exit time of this last region does not belong to L^q. For, if $n = 2$, the last region contains (1.5) with $\delta = \delta_{k-1}$, and if $n \geqq 3$, it contains

$$\{x \in \boldsymbol{R}^n: x_2^2 + x_3^2 > \delta_{k-1}^2\},$$

a region with an exit time having the same distribution as the exit time in the first example following Theorem 1.4.

The proof of Theorem 4.1 can now be completed by using the methods of Example 4.1. Note that, if $n \geqq 3$, one can even replace (4.3) by

$$\limsup_{x\to x_0;\, x\in R} E_x \log^+ \tau = \infty. \tag{4.6}$$

We now turn to some positive results that lead to the proof of Theorem 4.2.

LEMMA 4.1. *Let x_0 be a regular boundary point of R. If there is a ball B containing x_0 such that $\tau_{R\cup B}^{1/2} \in L^p$, then*

$$\lim_{x\to x_0;\, x\in R} E_x \tau^{p/2} = 0.$$

Here $\tau_{R\cup B}$ denotes the exit time of $R \cup B$. This suggests the question: Under what conditions does

$$\tau^{1/2} \in L^p \Rightarrow \tau_{R\cup B}^{1/2} \in L^p \tag{4.7}$$

for all balls B intersecting R? Although one may feel at first that (4.7) should always hold, the above examples reveal the contrary.

To prove Lemma 4.1, we need the following:

LEMMA 4.2. *Let x be a point in R whose distance to the boundary is at least δ. Then, for all y satisfying $|x - y| < \delta/3$ and all $\lambda > 0$,*

$$P_y(\tau > \lambda) \leqq 2^n P_x(\tau > \lambda/2). \tag{4.8}$$

PROOF. Let τ_k be the exit time of the ball B_k with center x and radius $k\delta/3$, $k = 1, 2, 3$, and let $y \in B_1$. Then

$$P_y(\tau > 2\lambda) \leqq P_y(\tau - \tau_2 > \lambda) + P_y(\tau_2 > \lambda)$$

and, by translation, the last probability satisfies

$$P_y(\tau_2 > \lambda) \leqq P_x(\tau_3 > \lambda) \leqq P_x(\tau > \lambda).$$

By the strong Markov property of X,

$$P_y(\tau - \tau_2 > \lambda) = E_y f(X_{\tau_2})$$

where $f(z) = P_z(\tau > \lambda)$. This expectation is the Poisson integral of f relative to B_2. The kernel at y is no larger than $a_n = 3 \cdot 2^{n-2}$ times the kernel at x. Therefore

$$E_y f(X_{\tau_2}) \leqq a_n E_x f(X_{\tau_2}) = a_n P_x(\tau - \tau_2 > \lambda) \leqq a_n P_x(\tau > \lambda).$$

These estimates imply (4.8).

PROOF OF LEMMA 4.1. Throughout the proof, suppose that $x \in R$ and $|x - x_0| < \delta/3$ where δ is the radius of B. By the regularity of x_0 (see **[8]**),

$$\lim_{x \to x_0} P_x(\tau > \lambda) = 0, \qquad \lambda > 0.$$

Therefore, by the Lebesgue dominated convergence theorem,

$$\lim_{x \to x_0} E_x \tau^{p/2} = \lim_{x \to x_0} \int_0^\infty (p/2)\, \lambda^{p/2-1}\, P_x(\tau > \lambda)\, d\lambda = 0$$

since, by Lemma 4.2,

$$P_x(\tau > \lambda) \leqq P_x(\tau_{R \cup B} > \lambda) \leqq 2^n\, P_{x_0}(\tau_{R \cup B} > \lambda/2)$$

and, by assumption, the last expression is integrable with respect to the measure $\lambda^{p/2-1}\, d\lambda$.

LEMMA 4.3. *If R is a region and B is a ball such that $\partial B \subset R$ and $\tau^{1/2} \in L^p$, then $\tau_{R \cup B}^{1/2} \in L^p$.*

PROOF. We shall show that $|x|^p$ is majorized by a superharmonic function in $R \cup B$. This implies, by a slight variation of the proof of Theorem 1.1, that $\tau_{R \cup B}^{1/2} \in L^p$.

Suppose that B has center x_0 and radius δ. Let B_0 be a slightly smaller ball with center x_0 and radius $\delta_0 < \delta$ such that $B - B_0 \subset R$. By Theorem 1.1, there is a function u harmonic in R majorizing $|x|^p$ in R. Let u_0 be the harmonic function in B with boundary values u on ∂B. If $n = 2$, let $U(x) = \log \delta - \log|x - x_0|$ and, if $n \geqq 3$, let $U(x) = |x|^{2-n} - \delta^{2-n}$. If $x \in R$ and $x \notin B_0$, let $u_1(x) = u(x) + MU(x)$ where the positive number M is chosen so that $u_1(x) \geqq u_0(x)$, $x \in \partial B_0$. Such a choice is possible since U is constant and positive on ∂B_0. Finally, let

$$\begin{aligned} v(x) &= u_0(x), \quad x \in B, \\ &= u_1(x), \quad x \notin B,\ x \in R. \end{aligned}$$

Note that v is continuous on $R \cup B$ since U vanishes on ∂B. Also, $u_1(x) \geqq u_0(x)$ on $\partial B_0 \cup \partial B$ implies, by the maximum principle, that the same inequality holds on $B - B_0$. Therefore, $u_1(x) \geqq v(x)$, $x \notin B_0$, $x \in R$. So each point of $R \cup B$ has a neighborhood on which v is the minimum of two harmonic functions. Therefore, v is superharmonic. If R is unbounded, then

$$\liminf_{|x| \to \infty;\, x \in R} \frac{v(x)}{|x|^p} \geqq 1.$$

Therefore, for some choice of positive numbers a and b, we have $|x|^p \leqq av(x) + b$, $x \in R \cup B$, which is the desired majorization.

This lemma can also be proved by using stopping times.

PROOF OF THEOREM 4.2. We can and do suppose that R contains the origin and, as in §3, let R_s denote the corresponding symmetrized region. There are two cases.

Case (i). Suppose some unbounded interval of the negative x_1-axis is disjoint from R_s. Let B be any ball intersecting R. Then $(R \cup B)_s$ also has this property: some translate of R_π (see the application of §1) contains $(R \cup B)_s$. Therefore, from (1.3) with $\alpha = \pi$ and (3.1) applied to $\tau_{R \cup B}$, we obtain $\tau_{R \cup B}^{1/2} \in L^p$, $0 < p < 1/2$. Lemma 4.1 now implies the theorem in this case.

Case (ii). Suppose no unbounded interval of the negative x_1-axis is disjoint from

R_s. Then $B_1 \subset B_2 \subset \cdots$ exist where B_k is a ball with center the origin, boundary $\partial B_k \subset R$, and $B_k \to \boldsymbol{R}^2$. In this case, Lemmas 4.1 and 4.3 imply the theorem.

5. Hardy spaces. Here let R be a region of the complex plane and τ the first time a complex Brownian motion starting at a point b in R leaves R. If F is a function analytic in the open unit disc D and p is a positive real number, let

$$\|F\|_{H^p} = \sup_{0<r<1} \left[\int_0^{2\pi} |F(re^{i\theta})|^p \, d\theta \right]^{1/p}.$$

THEOREM 5.1. *Suppose that F is analytic and univalent in D with range $F(D) = R$ and $F(0) = b$. Then*

$$c_p E_b(2\tau + |b|^2)^{p/2} \leqq \|F\|_{H^p}^p \leqq C_p E_b(2\tau + |b|^2)^{p/2}. \tag{5.1}$$

This is implied by Theorem 1.2 and shows the relevance of the powerful tool of conformal mapping for the study of exit times. In particular, $\tau^{1/2} \in L^p$ if and only if the conformal map $F \in H^p$.

Actually, the right-hand side of (5.1) holds without the assumption of univalence. The left-hand side can also be obtained under less restrictive conditions:

THEOREM 5.2. *Suppose that G is analytic in D with $G(D) \cap R$ nonempty and, for almost all θ, $\lim_{r\to 1} G(re^{i\theta})$ exists and belongs to the complement of R. Then*

$$G \in H^p \Rightarrow \tau^{1/2} \in L^p.$$

If, in addition, $G(0) = b \in R$, then

$$c_p E_b(2\tau + |b|^2)^{p/2} \leqq \|G\|_{H^p}^p.$$

Setting $R = F(D)$ and using the right-hand side of (5.1), we obtain

THEOREM 5.3. *Suppose that F and G are analytic in D with $F(D) \cap G(D)$ nonempty and, for almost all θ, $\lim_{r\to 1} G(re^{i\theta})$ exists and belongs to the complement of $F(D)$. Then*

$$G \in H^p \Rightarrow F \in H^p. \tag{5.2}$$

A special case, in which F is univalent, is implicit in the work of Hansen [**9**]. If, in addition, $F(0) = G(0)$, we also obtain the inequality

$$\|F\|_{H^p} \leqq c_p \|G\|_{H^p}$$

but this can be improved. Recently, Baernstein has shown that $c_p = 1$ (see [**4**]). Another proof can be constructed using covering maps as Hansen has recently discovered. Using a probability approach differing from the one above, we can give still another proof of $c_p = 1$. We summarize this result as follows.

THEOREM 5.4. *Suppose that F and G are analytic in D with $F(0) = G(0)$ and, for almost all θ, $\lim_{r\to 1} G(re^{i\theta})$ exists and belongs to the complement of $F(D)$. Then*

$$\|F\|_{H^p} \leqq \|G\|_{H^p}.$$

For further discussion, applications, and proofs see [**4**].

REFERENCES

1. A. Baernstein, *Integral means, univalent functions and circular symmetrization,* Acta Math. **133** (1974), 139–169.

2. A. Baernstein and B. A. Taylor, *Spherical rearrangements, subharmonic functions, and *-functions in n-space,* Duke Math. J. **43** (1976), 245–268.

3. D. L. Burkholder, *Distribution function inequalities for martingales,* Ann. Probability **1** (1973), 19–42. MR **51** #1944.

4. ———, *Exit times of Brownian motion, harmonic majorization, and Hardy spaces*, Advances in Math. (to appear).

5. D. L. Burkholder and R. F. Gundy, *Extrapolation and interpolation of quasi-linear operators on martingales,* Acta Math. **124** (1970), 249–304.

6. B. Dahlberg, *A note on sets of harmonic measure zero* (preprint).

7. J. L. Doob, *Semimartingales and subharmonic functions,* Trans. Amer. Math. Soc. **77** (1954), 86–121. MR **16**, 269.

8. E. B. Dynkin and A. A. Yushkevich, *Markov processes: Theorems and problems*, Plenum Press, New York, 1969. MR **39** #3585a.

9. L. J. Hansen, *Boundary values and mapping properties of H^p functions*, Math. Z. **128** (1972), 189–194. MR **47** #471.

UNIVERSITY OF ILLINOIS AT URBANA-CHAMPAIGN

Proceedings of Symposia in Pure Mathematics
Volume 31, 1977

A LIAPUNOV PRINCIPLE FOR SEMIMARTINGALES

HANS FÖLLMER

1. Introduction. Consider a system $\Sigma = (\Omega, \mathscr{F}, \mathscr{F}_t, P)$ where $(\Omega, \mathscr{F}, P)$ is a probability space and $(\mathscr{F}_t)_{t \geqq 0}$ a right-continuous family of σ-fields contained in $\mathscr{F}$. We want to discuss stochastic stability of subsets Λ of the product space $\Omega \times (0,\infty]$. Stochastic stability will be specified in various ways. It could mean, for example,

$$\int_0^\infty I_{\Lambda^c}(\cdot, t)\, dt < \infty \qquad P\text{-a.s.} \tag{1.1}$$

i.e., the system spends most of the time in Λ, or positive recurrence in the sense of

$$\liminf_{t \uparrow \infty} \frac{1}{t} \int_0^t I_\Lambda(\cdot, s)\, ds > 0 \qquad P\text{-a.s.} \tag{1.2}$$

i.e., the system spends at least a positive fraction of the time in Λ. Any local semimartingale $X = (X_t)_{t \geqq 0}$ over Σ, which behaves like a strict supermartingale outside of Λ, will be called a *Liapunov process* for Λ. A *Liapunov criterion* specifies further conditions on a Liapunov process which guarantee a given type of stochastic stability.

As an illustration consider the case where Σ is a representation of a nice Markov process $(\xi_t)_{t \geqq 0}$ with state space E and semigroup $(P_t)_{t \geqq 0}$, and where we are interested in stochastic stability of some subset A of E, for example in the sense of (1.1) or (1.2) with

$$\Lambda = \{(\omega, t) \,|\, \xi_t(\omega) \in A\}. \tag{1.3}$$

Let f be a function on E such that the process $X_t = f(\xi_t)$ $(t \geqq 0)$ is a local semi-

---

AMS (MOS) subject classifications (1970). Primary 60G45, 93E15.

martingale; this guarantees that

$$\mathscr{D}f \equiv \lim_{t \downarrow 0} \frac{1}{t}(f - P_t f)$$

exists up to a set of potential 0; cf. [11], [1]. Let us say that f is a Liapunov function for A if it is superharmonic outside of A with $\mathscr{D}f(x) > \varepsilon$ $(x \in E - A)$ for some $\varepsilon > 0$. In this case $X = (X_t)_{t \geqq 0}$ is a Liapunov process for Λ with

$$(\mathscr{D}X)(\omega, t) \equiv (\mathscr{D}f)(\xi_t(\omega)) > \varepsilon$$

outside of Λ. Under additional boundedness and regularity assumptions on f the set Λ is then positive recurrent in the sense of (1.2), or even stable in the sense of (1.1); cf., for example, [14]. Note that such Liapunov criteria are really statements on the sets $A_\varepsilon = \{\mathscr{D}f \leqq \varepsilon\}$, resp. $\Lambda_\varepsilon = \{\mathscr{D}X \leqq \varepsilon\}$, and thus express intrinsic properties of the function f, resp. of the local semimartingale X.

In our general context above this leads to the following *Liapunov principle*:

(1.4) *If X is a suitably "bounded" local semimartingale then the sets $\{\mathscr{D}X \leqq \varepsilon\}$ are "stable".*

In §3 we prove three versions of (1.4) where the "derivative" $\mathscr{D}X$ and the terms "bounded" and "stable" are specified with respect to some reference measure Q on $\Omega \times (0, \infty]$, which could be $Q = P \times dt$ as in (1.1) and (1.2). The arguments are based on the representation of semimartingales as signed measures on the σ-field of predictable sets; the needed facts are summarized in §2.

The problem of getting (1.2) in the case (1.3), but without assuming the Markov property for (ξ_t), came up in a discussion with W. Hildenbrand; cf. also [7]. Insofar as (1.4) extends some features of Markovian potential theory to a general martingale context, it may be viewed as a contribution to the program outlined in [4].

2. Semimartingales as signed measures. We assume that our system $\Sigma = (\Omega, \mathscr{F}, \mathscr{F}_t, P)$ has some "standard" properties as specified in [6, (1.4)]. Let $\bar{\Omega}$ denote the product space $\Omega \times (0, \infty]$ and $\mathscr{P}$ the σ-field of predictable sets in $\bar{\Omega}$; $\mathscr{P}$ is generated by all sets $A \times (t, \infty]$ with $t \geqq 0$ and $A \in \mathscr{F}_t$. Now consider a real-valued process $X = (X_t)_{t \geqq 0}$ adapted to $(\mathscr{F}_t)_{t \geqq 0}$ which is right-continuous pathwise and in L^1. X is called a *semimartingale* ([9], or *quasi-martingale* [5], or *F-process* [12]) if it has bounded conditional variation in the sense that

$$\text{(2.1)} \qquad \operatorname{var}(X) \equiv \sup \sum_{i=0}^{n} E[|X_{t_i} - E[X_{t_{i+1}} | \mathscr{F}_{t_i}]|] < \infty$$

where the supremum is taken over all finite sequences $0 = t_0 < \cdots < t_{n+1} = \infty$, and where we use the convention $X_\infty = 0$.

If X is a semimartingale then there exists a finite signed measure P^X on $\mathscr{P}$ with total mass $\|P^X\| = \operatorname{var}(X)$ such that

$$\text{(2.2)} \qquad P^X[A \times (t, \infty]] = E[X_t; A] \qquad (t \geqq 0, A \in \mathscr{F}_t).$$

For stopping times $S \leqq T$ and sets $A \in \mathscr{F}_S$ we have

$$\text{(2.3)} \qquad P^X[A \times (S, T]] = E[X_S - X_T; A]$$

where we write $A \times (S, T] = \{(\omega, t) | \omega \in A, S(\omega) < t \leqq T(\omega)\}$. (2.2) establishes a 1-1 correspondence between all semimartingales over Σ and all finite signed measures Q on $\bar{\Omega}$ which are *adapted* to Σ in the sense that they live on $\mathscr{P}$ and that for each $t \geqq 0$ the projection $Q_t[A] \equiv Q[A \times (t, \infty]]$ $(A \in \mathscr{F}_t)$ is absolutely continuous with respect to P.

A semimartingale X is a nonnegative supermartingale iff P^X is a nonnegative measure, a martingale iff P^X is concentrated on $\Omega \times \{\infty\}$, and a potential iff P^X is carried by $\Omega \times (0, \infty)$. Let us introduce the *lifetime* $\zeta((\omega, t)) \equiv t$, and the *explosion set*

$$C^X \equiv \{(\omega, t) | R_n(\omega) < t \ (n \geqq 1), \sup R_n(\omega) = t\}$$

where ζ is announced by the stopping times

$$R_n(\omega) = \inf\{t > 0 | X_t(\cdot) > n\}.$$

A potential X is a local martingale iff P^X is concentrated on C^X, and of class (D) iff $P^X[C^X] = 0$ or, alternatively, iff P^X vanishes on each evanescent set. A potential X is regular iff ζ is totally inaccessible with respect to P^X.

Let us now fix a measure $Q \geqq 0$ which is adapted to Σ and finite on each strip $\Omega \times (0, t)$, and let us introduce the *derivative* of a semimartingale X as $\mathscr{D}X = dP^X/dQ$, the Radon-Niklodým derivative of the absolutely continuous part of P^X with respect to Q. $\mathscr{D}X$ is thus a predictable process, and it may be identified in terms of a concrete limit procedure which depends on the structure of Q. If, for example, $Q = P^Y$ with a local martingale $Y = (Y_t)_{t \geqq 0}$ then we can identify $\mathscr{D}X$ in terms of a *Fatou theorem*:

$$\mathscr{D}X(\omega, t) = \lim_{s \uparrow t} \frac{X_s(\omega)}{Y_s(\omega)} \qquad Q\text{-a.s.} \tag{2.4}$$

If, on the other hand, $Q = P^Y$ with a regular potential Y then $\mathscr{D}$ may be viewed as an *infinitesimal generator*:

$$\mathscr{D}X(\omega, t) = \lim_{h \downarrow 0} \frac{E[X_t - X_{t+h} | \mathscr{F}_t]}{E[Y_t - Y_{t+h} | \mathscr{F}_t]}(\omega) \qquad Q\text{-a.s.} \tag{2.5}$$

if the versions on the right are properly chosen.

For proofs and further details we refer to **[1]**, **[6]**, **[13]**.

3. Three versions of the Liapunov principle. We are now going to specify with respect to Q the loose terms in (1.4). Let us put

$$\Lambda = \Lambda_\varepsilon = \{\mathscr{D}X \leqq \varepsilon\}$$

for some $\varepsilon > 0$. The aim is to show that under suitable "bounds" on X the predictable set Λ is "substantial" for Q. Note that the assumption, that X is a semimartingale, means a bound on the conditional variation of X.

We can always split Q into an *explosive* part, which is carried by some evanescent set, and into a *tame* part which vanishes on each evanescent set (same argument as in **[6**, 2.6**]**). Let us therefore argue piecewise.

Assume first that Q is explosive, and define $X_-(\omega, t) = \limsup_{s \uparrow t} X_s(\omega)$.

(3.1) VERSION I. *If X is a semimartingale and $X_- < \infty$ Q-a.s. then Q is carried by Λ.*

PROOF. We may assume that Q is finite, and thus of the form $Q = P^Y$ for some supermartingale $\geqq 0$. Since Q is carried by some evanescent set, Y is in fact a local martingale [6, 2.6], and due to (2.4) we have

$$\mathscr{D}X(\omega, t) = \lim_{s \uparrow t} \frac{X_s(\omega)}{Y_s(\omega)} \qquad Q\text{-a.s.}$$

But $\lim_{s \uparrow t} Y_s(\omega) = +\infty$ Q-a.s. since Q is carried by the explosion set C^Y, and thus $X_- < +\infty$ implies $-\infty \leqq \mathscr{D}X \leqq 0$ Q-a.s.

From now on we assume that Q is tame. This allows us to view Q as a measure on $\bar{\mathscr{F}}$, the product of $\mathscr{F}$ and the Borel σ-field on $(0, \infty]$ (extension by predictable projection, since two predictable projections of a set $A \in \bar{\mathscr{F}}$ differ at most by an evanescent set). Thus Q is of the form $Q = P \times dB$ for some right-continuous increasing process $B = (B_t)_{t \geqq 0}$ and, due to the definition of Q via predictable projection, B is in fact a predictable process. This is essentially Doléans-Dade's construction of the increasing process associated to a potential of class (D); cf. [3], [6] for details. Since Q is finite on $\Omega \times (0, t]$ the random variables B_t are integrable.

(3.2) VERSION II. *If X is a semimartingale then we have*

$$Q[\Lambda^c] = E\left[\int_0^\infty I_{\Lambda^c}(\cdot, t)\, dB_t\right] < \infty.$$

(3.3) REMARK. In particular we have

$$\int_0^\infty I_{\Lambda^c}(\cdot, t)\, dB_t < \infty \qquad P\text{-a.s.}$$

which means that the system spends most of the time in Λ if there is time enough, i.e., on the set $\{B_\infty = \infty\}$. Note that in Version II we could replace Λ by $\{|\mathscr{D}X| \leqq \varepsilon\}$.

PROOF.

$$Q[\Lambda^c] \leqq Q[|\mathscr{D}X| > \varepsilon] \leqq \frac{1}{\varepsilon}\int |\mathscr{D}X|\, dQ \leqq \frac{1}{\varepsilon}\, \|P^X\| < \infty.$$

Let us now consider the less immediate case where $X = (X_t)_{t \geqq 0}$ is a *local* semimartingale. We are thus dropping the bound on the total variation of X. Let us in fact assume that for each $t_0 > 0$

(3.4) $$(X_{t \wedge t_0})_{t \geqq 0} \text{ is a semimartingale of class (D).}$$

This allows us to define P^X consistently on the strips $\Omega \times (0, t)$, and so we have the derivative $\mathscr{D}X = dP^X/dQ$ on $\Omega \times (0, \infty)$. Note also that under (3.4) the Doob decomposition of X takes the form

(3.5) $$X_t = M_t - A_t$$

where $M = (M_t)_{t \geqq 0}$ is a martingale (not only a local martingale) and $A = (A_t)$ is a process with paths of bounded variation.

(3.6) REMARK. The assumption in (3.4), that X can be localized by constant stopping times, is more than we really need. But it does simplify the discussion, because it allows us to introduce the "usual conditions", i.e., to replace $(\mathscr{F}_t)$ by its right-continuous completion with respect to P; cf. [2, pp. 182–183]. From now on $(\mathscr{F}_t)$ will denote that completion, and $\mathscr{P}$ the corresponding σ-field of predictable sets. (3.4) implies in fact that P^X may be viewed, at least on each strip $\Omega \times (0, t)$, as a measure on the (extended) σ-field $\mathscr{P}$; cf. [6, 2.6].

Let us now look at the set

$$\Lambda = \Lambda_\varepsilon = \{\mathscr{D}X \leqq \varepsilon\} \cap \Omega \times (0, \infty).$$

We assume from now on that X is *nonnegative*, and that it *behaves like a supermartingale outside of* Λ in the sense that $P^X \geqq 0$ on $\Lambda^c \cap \Omega \times (0, \infty)$. Moreover we assume $M_t \in L^2$ $(t \geqq 0)$ for the martingale part in (3.5) and denote by $\langle M, M\rangle = (\langle M, M\rangle_t)_{t\geqq 0}$ the predictable increasing process associated to the submartingale $(M_t^2)_{t\geqq 0}$. Our aim is *positive recurrence* of Λ in the sense of

$$\liminf_{t\uparrow\infty} \frac{1}{B_t}\int_0^t I_\Lambda(\cdot, s)\, dB_s > 0 \qquad P\text{-a.s.}$$

In view of immediate counterexamples like $B_t = t$ and $X_t = 1 - (t - n)$ on $[n, n + 1)$, or $X_t = 2^{n+1} - 1 - t$ on $[2^n, 2^{n+1} - 1)$ and $= (2^{n+1} - 1)(t - 2^{n+1} + 1)$ on $[2^{n+1} - 1, 2^{n+1})$, we need some further restrictions. Let us say that a set $H(\Lambda) \in \mathscr{P}$ is a *hull* of Λ if, starting from Λ, the time needed to get out of $H(\Lambda)$ is bounded away from 0, i.e., if there is some $\gamma > 0$ such that

$$E[B_{D_{H(\Lambda)^c}\circ\theta_S} - B_S | \mathscr{F}_S] \geqq \gamma \quad \text{on } \{D_\Lambda \circ \theta_S = 0\} \tag{3.7}$$

for any stopping time S. Here and in the sequel we denote by $D_A \circ \theta_S = \inf\{t \geqq S | (\cdot, t) \in A\}$ the first entrance time into a set $A \in \mathscr{P}$, counting from time S on.

(3.8) REMARKS. (1) For $B_t = t$ and $\gamma > 0$ the set $\{(\omega, t + s) | (\omega, t) \in \Lambda, 0 \leqq s \leqq \gamma\}$ is a hull of Λ.

(2) If we put a bound on the "second derivative" and assume that the predictable process $Y = \mathscr{D}X$ is in fact a local semimartingale with $P^X \ll Q$ and density $\mathscr{D}^{(2)}X = \mathscr{D}Y$ bounded from below on Λ, then it is easy to show that Λ_ε is a hull for Λ_{ε_0} $(0 < \varepsilon_0 < \varepsilon)$. In this case our Version III below, applied to Λ_{ε_0}, yields positive recurrence of the set $\Lambda = \Lambda_\varepsilon$ itself.

Let us fix a hull $H(\Lambda)$ and let us define $R_0 = T_0 = 0$ and

$$T_n \equiv D_\Lambda \circ \theta_{S_n}, \qquad R_n \equiv D_{H(\Lambda)^c} \circ \theta_{S_n+T_n} \qquad (n \geqq 1)$$

where the times $S_n \equiv \sum_{i=0}^{n-1} (T_i + R_i)$ are the successive exit times from $H(\Lambda)$ via Λ. We can now state

(3.9) VERSION III. *Suppose*

(i) $\sup_n X_{S_n} < \infty$ *P-a.s.*

(ii) $\langle M, M\rangle \ll B$ *with density bounded by some constant c (at least outside of* Λ*). Then we have*

$$\liminf_{t\uparrow\infty} \frac{1}{B_t}\int_0^t I_{H(\Lambda)}(\cdot, s)\, dB_s > 0 \tag{3.10}$$

P-a.s. on $\{B_\infty = \infty\}$.

(3.11) REMARK. Assumption (i) means that X is "bounded near Λ", and (ii) is essentially a bound on the conditional variance of X.

The increase in B during the time spent in Λ^c between the nth exit from $H(\Lambda)$ and the next return to Λ is given by

$$\tau_n \equiv B_{(S_n+T_n)-} - B_{S_n} + \Delta B_{S_n+T_n} I_{\Lambda^c}(\cdot, S_n + T_n)$$

(B_{t-} denotes the left limit and $\Delta B_t = B_t - B_{t-}$ the jump in t). Let us also put

$$\rho_n \equiv B_{S_{n+1}} - (B_{S_n} + \tau_n), \qquad \sigma_n \equiv B_{S_n} = \sum_{i=1}^{n-1} (\tau_i + \rho_i)$$

(with $\tau_n = \rho_n = 0$ on $\{S_n = \infty\}$).

(3.12) LEMMA. *For* $p = 1, 2$ *we have*

$$\beta_p \equiv \sup_n E[\tau_n^p | \mathscr{F}_{S_n}] < \infty \qquad P\text{-a.s.}$$

PROOF. We fix $n \geqq 1$ and write $S = S_n$, $T = T_n$, $\tau = \tau_n$.

(1) The graph $[T]$ of T satisfies $[T] \cap \Lambda = \Lambda - \Lambda \cap \{\zeta > T\} \in \mathscr{P}$, and so the random variable $\tilde{T}$ with graph $[\tilde{T}] = [T] \cap \Lambda + L^c \times \{\infty\}$, where $L = I\{_\Lambda(\cdot, T(\cdot)) = 1\} \in \mathscr{F}_{T-}$ due to [**2**, IV.67], is predictable; cf. [**2**, IV.87 and the "Corrections"]. Announce $\tilde{T}$ by stopping times $(\tilde{U}_k)_{k\geqq 1}$; cf. [**2**, IV.77]. Then the stopping times $U_k = \tilde{U}_k \wedge T$ increase to T and announce T on the set L. In particular we have

$$\bigcup_k (S, U_k] = (S, T] \cap \Lambda^c. \tag{3.13}$$

(2) Let us show (3.12) for $p = 1$. Using $P^X \geqq 0$ on Λ^c and (3.13) we obtain

$$\begin{aligned} \varepsilon E[\tau; A] &\leqq \int_{A\times(S,T]\cap\Lambda^c} \mathscr{D}X\, dQ \leqq P^X[A \times (S, T] \cap \Lambda^c] \\ &= \lim_k P^X[A \times (S, U_k]] = \lim_k E[X_S - X_{U_k}; A \cap \{U_k > S\}] \\ &\leqq E[X_S; A] \qquad (A \in \mathscr{F}_S), \end{aligned}$$

and in particular

$$E[\tau | \mathscr{F}_S] \leqq \frac{1}{\varepsilon} \sup_n X_{S_n} < \infty.$$

(3) In order to obtain (3.12) for $p = 2$ we consider the nonnegative process $Y_t = X_{s+t} I_{\{t<T\}}$ $(t \geqq 0)$. $Y = (Y_t)_{t\geqq 0}$ is adapted to $\mathscr{G}_t = \mathscr{F}_{S+t}$ $(t \geqq 0)$, and in fact a supermartingale:

$$\begin{aligned} E[Y_s - Y_t; A] &\geqq E[Y_s - \liminf Y_{(U_k-s)\wedge t}; A] \\ &\geqq \liminf E[X_{S+s} - X_{U_k\wedge(S+t)}; A \cap \{U_k > S + s\}] \\ &= \liminf P^X[(A \cap \{U_k > S + s\}) \times (S + s, U_k \wedge (S + t)]] \\ &\geqq 0 \qquad (A \in \mathscr{G}_s) \end{aligned}$$

since $P^X \geqq 0$ on $\Lambda^c \supseteq \bigcup_k (S, U_k]$. Let us write $Y_t = \tilde{M}_t - \tilde{A}_t$ with

$$\begin{aligned} \tilde{M}_t &= M_{S_n+t} I_{\{t<T\}} + M_{(S_n+T)-} I_{\{T\leqq t\}}, \\ \tilde{A}_t &= A_{S_n+t} I_{\{t>T\}} + M_{(S_n+T)-} I_{\{T\leqq t\}}. \end{aligned}$$

$(\tilde{A}_t - \tilde{A}_0)_{t\geqq 0}$ is an increasing right-continuous process which generates (Y_t). By

[8, VII.60(b)] the *predictable* increasing process (C_t) associated to (Y_t) satisfies

$$\begin{aligned} E[C_\infty^2 | \mathscr{G}_0] &\leqq 8E[(\tilde{A}_\infty - \tilde{A}_0)^2 | \mathscr{G}_0] \\ &= 8E[(M_{(S+T)-} - M_S + X_S)^2 | \mathscr{G}_0] \\ &= 8E[(M_{(S+T)-} - M_S)^2 | \mathscr{G}_0] + 8X_S^2 \\ &\leqq 8[cE[\tau | \mathscr{G}_0] + X_S^2] \end{aligned}$$

due to assumption (ii). On the other hand we have $C_\infty \geqq A_T \geqq \varepsilon\tau$, and so we obtain

$$\varepsilon^2 E[\tau^2 | \mathscr{G}_0] \leqq 8\left[c\beta_1(\cdot) + \left(\sup_n X_{S_n}\right)^2\right].$$

We are now ready for the

PROOF OF VERSION III. Due to (3.12) we have $\tau_n < \infty$ P-a.s. on $\{S_n < \infty\}$, and so it is no loss of generality to assume $S_n < \infty$ for all $n \geqq 1$.

(1) By (3.12) we have

$$\tilde{\tau}_n \equiv E[\tau_n | \mathscr{F}_{S_n}] \leqq \beta_1, \qquad V_n \equiv \sum_{k=1}^n E[\tau_k^2 | \mathscr{F}_{S_k}] \leqq n\beta_2,$$

hence

$$\begin{aligned} \limsup &\left[\frac{\tau_1 + \cdots + \tau_n}{n} - \beta_1\right] \\ &\leqq \limsup \frac{\tau_1 + \cdots + \tau_n - (\tilde{\tau}_1 + \cdots + \tilde{\tau}_n)}{n} \\ &\leqq \beta_2 \limsup \frac{\tau_1 + \cdots + \tau_n - (\tilde{\tau}_1 + \cdots + \tilde{\tau}_n)}{V_n}. \end{aligned}$$

But by a law of large numbers of Neveu-Dubins-Freedman (cf., for example, [9, II.65]) the second factor on the right is actually P-a.s. a finite limit which is equal to 0 on $\{V_\infty = \infty\}$. In any event we may conclude

(3.14) $$\limsup \frac{\tau_1 + \cdots + \tau_n}{n} \leqq \beta_1 \qquad P\text{-a.s.}$$

(2) By definition of a hull we have

$$\tilde{\rho}_n \equiv E[\rho_n | \mathscr{F}_{S_n+T_n}] \geqq \gamma$$

for some $\gamma > 0$. Take $0 < \delta < \gamma$ and choose c_n such that $\rho_n' \equiv \rho_n \wedge c_n$ satisfies $E[\rho_n' | \mathscr{F}_{S_n+T_n}] \geqq \delta$. Note

$$\delta^2 \leqq E[\rho_n'^2 | \mathscr{F}_{S_n+T_n}] < \infty$$

and apply the law of large numbers as in (a). This yields

$$\liminf \frac{\rho_1 + \cdots + \rho_n}{n} \geqq \liminf \frac{\rho_1' + \cdots + \rho_n'}{n} \geqq \delta;$$

hence

(3.15) $$\liminf \frac{\rho_1 + \cdots + \rho_n}{n} \geqq \gamma.$$

(3) On $\{S_n \leqq t < S_{n+1}\}$ we have

$$\frac{1}{B_t}\int_0^t I_{H(\Lambda)^c}\, dB_s \leqq \frac{\tau_1 + \cdots + \tau_n}{\sigma_n}$$

where $\tau_{n+1}/\sigma_{n+1} \leqq (\tau_{n+1}/n)\, n/(\rho_1 + \cdots + \rho_n)$ goes to 0 due to (3.14) and (3.15). Thus

$$\limsup_{t\uparrow\infty} \frac{1}{B_t}\int_0^t I_{H(\Lambda)^c}\, dB_s \leqq \limsup_{n\uparrow\infty} \frac{\tau_1 + \cdots + \tau_n}{\tau_1 + \cdots + \tau_n + \rho_1 + \cdots + \rho_n} \leqq \frac{\beta_1}{\beta_1 + \gamma}$$

by (3.14) and (3.15), and

$$\liminf_{t\uparrow\infty} \frac{1}{B_t}\int_0^t I_{H(\Lambda)}\, dB_s \geqq \frac{\gamma}{\beta_1 + \gamma} > 0.$$

REFERENCES

1. H. Airault and H. Föllmer, *Relative densities of semimartingales*, Invent. Math. **27** (1974), 299–327. MR **51** #1976.
2. C. Dellacherie and P.-A. Meyer, *Probabilités et potentiel,* rev. ed., Publ. Inst. Math. Univ. Strasbourg XV, Hermann, Paris, 1975.
3. C. Doléans-Dade, *Existence du processus croissant naturel associé à un potentiel de la classe* (D), Z. Wahrscheinlichkeitstheorie und Verw. Gebiete **9** (1968), 309–314. MR **39** #7667.
4. J. L. Doob, *Martingale theory-potential theory*, Potential Theory (C.I.M.E. I Ciclo, Stresa, 1969), Edizioni Cremonese, Rome, 1970, pp. 203–206. MR **42** #6261.
5. D. L. Fisk, *Quasi-martingales*, Trans. Amer. Math. Soc. **120** (1965), 369–389. MR **33** #767.
6. H. Föllmer, *On the representation of semimartingales*, Ann. Probability **1** (1973), 580–589. MR **50** #5929.
7. J. Lamperti, *Criteria for the recurrence of transience of stochastic process.* I, J. Math. Anal. Appl. **1** (1960), 314–330. MR **23** #A4166.
8. P.-A. Meyer, *Probability and potentials*, Blaisdell, Waltham, Mass., 1966. MR **34** #5118; 5119.
9. ———, *Intégrales stochastiques.* I, II, III, IV, Séminaire de Probabilités (Univ. Strasbourg, Strasbourg, 1966/67), Vol. I, Springer-Verlag, Berlin, 1967, pp. 72–94, 95–117, 118–141, 142–162. MR **37** #7000.
10. ———, *Martingales and stochastic integrals.* I, Lecture Notes in Math., vol. 285, Springer-Verlag, Berlin and New York, 1972.
11. G. Mokobodzki, *Densité relative de deux potentiels comparables*, Séminaire de Probabilités. IV (Univ. Strasbourg, 1968/69), Lecture Notes in Math., vol. 124, Springer-Verlag, Berlin, 1970, pp. 170–194. MR **45** #3747.
12. S. Orey, *F-processes*, Proc. Fifth Berkeley Sympos. Math. Statist. and Probability (Berkeley, Calif., 1965/66), Vol. II: Contributions to Probability Theory, Part 1, Univ. of California Press, Berkeley, Calif., 1967, pp. 301–313. MR **35** #4975.
13. C. Stricker, *Mesure de Föllmer en theorie des quasi-martingales*, Séminaire de Probabilités. IX (Univ. of Strasbourg, 1973/74 and 1974/75), Part 2, Lecture Notes in Math., vol. 465, Springer-Verlag, Berlin and New York, 1975, pp. 408–419. MR **51** #9136.
14. W. M. Wonham, *Liapunov criteria for weak stochastic stability*, J. Differential Equations **2** (1966), 195–207. MR **33** #2906.

UNIVERSITÄT BONN

Proceedings of Symposia in Pure Mathematics
Volume 31, 1977

APPLICATIONS OF DUAL PROCESSES TO DIFFUSION THEORY

R. HOLLEY*, D. STROOCK* AND D. WILLIAMS

Introduction. The study of certain infinite systems of interacting processes can be viewed as the investigation of elliptic operators on infinite-dimensional spaces. For instance, spin-flip models involve looking at operators

$$\mathscr{L} = \sum_{k\in S} c_k(\eta)\Delta_k,^1$$

on $E = (\{-1, 1\})^S$, where S is a countable, infinite set. The central problem to be solved is the determination of the ergodic, or lack of ergodic, properties of the semigroup $e^{t\mathscr{L}}$. The analytic difficulties encountered are easy to appreciate even in the simplest cases. For instance, suppose $c_k(\cdot) \equiv 1$ for all $k \in S$. Then $e^{t\mathscr{L}}f(\eta) = \int P(t, \eta, d\xi)f(\xi)$, where

$$P(t, \eta, \cdot) = \prod_{k\in S}\left(\frac{1+e^{-2t}}{2}\delta_{\{\eta_k\}} + \frac{1-e^{-2t}}{2}\delta_{\{-\eta_k\}}\right).$$

Thus the transition function does not admit a reference measure. In fact, $P(s, \eta, \cdot) \perp P(t, \eta, \cdot)$ for $s \neq t$; even worse, $P(t, \eta, \cdot) \perp P(t, \eta', \cdot)$ if $\eta_k \neq \eta'_k$ for infinitely many $k \in S$. Finally, $e^{t\mathscr{L}}f(\eta) \to \int f\, d\mu_0$ as $t \uparrow \infty$ for $f \in C(E)$, where $\mu_0 = \prod_{k\in S}(\frac{1}{2}\delta_{\{-1\}} + \frac{1}{2}\delta_{\{1\}})$, but $P(t, \eta, \cdot) \perp \mu_0$ for all $t > 0$. Thus, most of the standard results about the ergodic properties of Markov processes are not applicable to this situation.

From the analytic point of view, the origin of these difficulties is the lack of a priori estimates for elliptic operators on infinite-dimensional spaces. Thus techniques have to be developed which do not rely on such estimates. One technique which turns out to be useful, in spite of its serious limitations, is that of "dual

AMS (*MOS*) *subject classifications* (1970). Primary 60J60; Secondary 60K35.
*Research partially supported by N.S.F. MPS74–18926.
[1] Δ_k is the partial difference operator with respect to coordinate k.

processes" (cf. [3]). For the benefit of those not familiar with infinite systems of interacting processes, we will describe this technique in the context of diffusions on a finite-dimensional torus. It is not our contention that dual processes are the only, or even the best, way to study such diffusions; the numerous analytic results known about elliptic operators in finite-dimensional spaces are undoubtedly more powerful tools. What we are trying to advertise is a technique which is insensitive to dimension and therefore is applicable to infinite-dimensional problems. The ability of this technique to handle situations in which a priori estimates are absent is manifested below in the ease with which it deals with degenerate operators.

1. Dual processes and the question of uniqueness. Let T^d be the d-dimensional torus and use Δ to stand for the Laplacian on T^d. Given a continuous nonnegative function $a(\vec{\theta})$ on T^d, consider the elliptic operator $\mathscr{L} = a(\vec{\theta})\Delta$. We will say that the semigroup $\{T_t: t \geqq 0\}$ on $B(T^d)$ is a *Markov version of* $e^{t\mathscr{L}}$ if T_t is a nonnegative conservative, contraction Markov semigroup and $T_t f - f = \int_0^t T_s \mathscr{L} f\, ds$ for $f \in C^2(T^d)$. The question of uniqueness which we wish to discuss is whether there is more than one such $\{T_t: t \geqq 0\}$ (that there is at least one can be seen from the work of Krylov [5]). In order to apply dual processes to this problem, we will assume that

$$a(\vec{\theta}) = \gamma - \varphi(\vec{\theta}), \tag{1}$$

where $\gamma \geqq 1$ and

$$\varphi(\theta) = \sum_{\vec{n} \in Z^d}^{\infty} p_{\vec{n}}\, e^{i\vec{n}\cdot\vec{\theta}}, \tag{2}$$

where $\{p_{\vec{n}}: \vec{n} \in Z^d\}$ is a symmetric probability measure on Z^d. (This assumption is stronger than absolutely necessary but it facilitates the presentation.) Let $\mathscr{A}$ denote the generator of the symmetric random walk on Z^d determined by $\{p_{\vec{n}}: \vec{n} \in Z^d\}$. Then an easy computation yields:

$$\mathscr{L} e_{\vec{n}}(\vec{\theta}) = |\vec{n}|^2 \mathscr{A} e_{\vec{\theta}}(\vec{n}) - (\gamma - 1)|\vec{n}|^2 e_{\vec{\theta}}(\vec{n}), \tag{3}$$

where $e_{\vec{n}}(\vec{\theta}) = e_{\vec{\theta}}(\vec{n}) = e(\vec{n}, \vec{\theta}) \equiv e^{i\vec{n}\cdot\vec{\theta}}$. Thus if $\{T_t: t \geqq 0\}$ is any version of $e^{t\mathscr{L}}$ and $u_{\vec{\theta}}(t, \vec{n}) = T_t e_{\vec{n}}(\theta)$, then

$$(\partial u_{\vec{\theta}}/\partial t)(t, \vec{n}) = T_t \mathscr{L} e_{\vec{n}}(\theta) = |\vec{n}|^2 \mathscr{A} u_{\vec{\theta}}(t, \vec{n}) - (\gamma - 1)|\vec{n}|^2 u_{\vec{\theta}}(t, \vec{n}). \tag{4}$$

Given $N \geqq 1$, let $\{\hat{P}^{(N)}_{\vec{n}}: \vec{n} \in Z^d\}$ be the Markov family on $D([0, \infty), Z^d)$ generated by $\mathscr{X}_{[0,N)}(|\vec{n}|)\, |\vec{n}|^2 \mathscr{A}$ and set $\zeta^{(N)} = \inf\{t \geqq 0: |\vec{n}(t)| \geqq N\}$. Then $\hat{P}^{(N+1)}_{\vec{n}}$ equals $\hat{P}^{(N)}_{\vec{n}}$ up until $\zeta^{(N)}$, and by (4)

$$\begin{aligned} T_t e_{\vec{n}}(\vec{\theta}) = {} & \hat{E}^{(N)}_{\vec{n}}\left[e_{\vec{\theta}}(\vec{n}(t)) \exp\left(-(\gamma - 1)\int_0^t |\vec{n}(s)|^2\, ds\right), \zeta^{(N)} > t\right] \\ & + \hat{E}^{(N)}_{\vec{n}}\left[u_{\vec{\theta}}(t - \zeta^{(N)}, \vec{n}(\zeta^{(N)})) \exp\left(-(\gamma - 1)\int_0^{\zeta^{(N)}} |\vec{n}(s)|^2\, ds\right), \zeta^{(N)} \leqq t\right]. \end{aligned} \tag{5}$$

Here $\hat{E}^{(N)}_{\vec{n}}[\cdot]$ denotes expectation value with respect to $\hat{P}^{(N)}_{\vec{n}}$ and we have used the Feynman-Kac formula. Given a probability measure μ on T^d, define $\hat{\mu}(\vec{n}) = \int_{T^d} e_{\vec{n}}(\vec{\theta})_{\mu}(d\vec{\theta})$. Then (5) implies

$$\begin{aligned} \widehat{T_t^* \mu}(\vec{n}) = {} & \hat{E}^{(N)}_{\vec{n}}\left[\hat{\mu}(\vec{n}(t)) \exp\left(-(\gamma - 1)\int_0^t |\vec{n}(s)|^2\, ds\right), \zeta^{(N)} > t\right] \\ & + \hat{E}^{(N)}_{\vec{n}}\left[\widehat{T^*_{t-\zeta^{(N)}}\mu}(\vec{n}(\zeta^{(N)})) \exp\left(-(\gamma - 1)\int_0^{\zeta^{(N)}} |\vec{n}(s)|^2\, ds\right), \zeta^{(N)} \leqq t\right], \end{aligned} \tag{6}$$

where $\{T_t^*: t \geqq 0\}$ is the adjoint semigroup to $\{T_t: t \geqq 0\}$. If

(7) $$\lim_{N\to\infty} \hat{E}_{\vec{n}}^{(N)}\left[\exp\left(-(\gamma-1)\int_0^{\zeta^{(N)}} |\vec{n}(s)|^2\, ds\right), \zeta^{(N)} \leqq t\right] = 0, \quad \vec{n} \in Z^d \text{ and } t > 0,$$

then (6) implies that T_t^*, and therefore T_t is uniquely determined. Clearly (7) will hold if $\gamma > 1$. The uniqueness of $\{T_t: t \geqq 0\}$ when $\gamma > 1$ can, of course, be obtained in other ways by taking advantage of the strict ellipticity of $\mathscr{L}$ (cf. [6]). Thus we will restrict our attention to the case $\gamma = 1$, so that (7) becomes

(8) $$\lim_{N\to\infty} \hat{P}_{\vec{n}}^{(N)}(\zeta^{(N)} \leqq t) = 0, \qquad \vec{n} \in Z^d \text{ and } t > 0.$$

Next observe that (8) obtains if the random walk generated by $\mathscr{A}$ is recurrent. Indeed, the process $\vec{n}(t)$ is simply a random time change of this random walk until it hits 0, at which time it is absorbed. Since the random walk is symmetric, recurrence is equivalent to

(9) $$\int_{T^d} \frac{1}{1-\varphi(\vec{\theta})}\, d\vec{\theta} = \infty.$$

Thus there is only one Markov version of $\exp(t(1-\varphi(\vec{\theta}))\Delta)$ if (9) holds. When $d = 1$, the condition (9) is equivalent to (8). In fact, if (9) fails, then

$$\sup_N P_n^{(N)}((\exists\, t) n(t) = 0) < 1 \quad \text{and} \quad \sup_N E_n^{(N)}\left[\int_0^\infty \mathscr{X}_{(0,N)}(|\vec{n}(t)|)\, dt\right] < \infty$$

for all $n \neq 0$. The fact that (9) guarantees uniqueness when $d = 1$ is a special case of a result due to Watanabe and Yamada [7].

For $d \geqq 2$, (9) is much stronger than (8). For instance, (8) is implied by the condition

(10) $$\sum_{\vec{n}\in Z^d} |\vec{n}|^2 p_{\vec{n}} < \infty.$$

To see this, note that (10) implies that $\sup_N E_{\vec{n}}^{(N)}[|\vec{n}(t)|^2] < \infty$. That uniqueness is implied by (10) can also be concluded from Itô's criterion for the uniqueness solutions to stochastic differential equations.

An example to which the dual process technique applies and other techniques fail is obtained by taking $d = 2$ and choosing the $p_{\vec{n}}$ so that (10) fails and yet (9) holds. Such an example was produced by Chung and Fuchs [1].

It is interesting to see what this line of reasoning yields when there is more than one version of $e^{t\mathscr{L}}$. Suppose $\mathscr{L} = (1-\varphi(\theta))\Delta$, where φ is as in (2). A *Feller version of* $e^{t\mathscr{L}}$ is a Markov version which takes $C(T^d)$ into itself. Let $M^+(T^d)$ be the set of nonnegative measures on T^d and set $M(T^d) = M^+(T^d) - M^+(T^d)$. If $\mu \in M(T^d)$, use $\|\mu\|$ to denote the variation norm of μ and let $\hat{\mu}: Z^d \to \boldsymbol{C}$ be defined by $\hat{\mu}(\vec{n}) = \int e_n(\vec{\theta})\mu(d\vec{\theta})$. Let $\widehat{M^+(T^d)}$ and $\hat{M}(T^d)$ be, respectively, the images of $M^+(T^d)$ and $M(T^d)$ under $\hat{}$ and define $\|\hat{\mu}\| = \|\mu\|$ for $\hat{\mu} \in \hat{M}(T^d)$. Clearly $\hat{M}(T^d)$ with $\|\cdot\|$ is a Banach space. Given $\hat{\mu} \in \hat{M}(T^d)$, define $\hat{\mathscr{L}}\mu(\vec{n}) = |\vec{n}|^2 \mathscr{A}\hat{\mu}(\vec{n})$, $\vec{n} \in Z^d$, where $\mathscr{A}\hat{\mu}(\vec{n}) \cdot \sum_{\vec{k}} p_{\vec{k}}(\hat{\mu}(\vec{n}+\vec{k}) - \hat{\mu}(\vec{n}))$, and note that $|\hat{\mathscr{L}}\hat{\mu}(\vec{n})| \leqq |\vec{n}|^2\|\hat{\mu}\|$. We will say that $\{\hat{T}_t: t \geqq 0\}$ is a *Bochner version* of $e^{t\hat{\mathscr{L}}}$ if $\{\hat{T}_t: t \geqq 0\}$ is a semigroup of contractions on $\hat{M}(T^d)$ into itself such that $\widehat{M^+(T^d)}$ is invariant under $\hat{T}_t$ for all $t \geqq 0$, $\hat{T}_t\hat{\mu}_m(\vec{n}) \to \hat{T}_t\hat{\mu}(\vec{n})$ for all $t \geqq 0$ and $\vec{n} \in Z^d$ as $m \to \infty$ if $\{\hat{\mu}_m\}_1^\infty$ is a bounded sequence in $\hat{M}(T)$ such that $\hat{\mu}_m(\vec{n}) \to \hat{\mu}(\vec{n})$ as $m \to \infty$ for all $\vec{n} \in Z$, and

(11) $$\hat{T}_t\hat{\mu}(\vec{n}) - \hat{\mu}(\vec{n}) = \int_0^t \hat{\mathscr{L}}\hat{T}_s\hat{\mu}(\vec{n})\, ds \quad \text{for } t \geqq 0, \vec{n} \in Z^d, \text{ and } \hat{\mu} \in \hat{M}(T^d).$$

THEOREM. *If* $\{T_t: t \geqq 0\}$ *is a Feller version of* $e^{t\mathscr{L}}$ *and* $\{\hat{T}_t: t \geqq 0\}$ *is defined on* $\hat{M}(T^d)$ *by* $\hat{T}_t\hat{\mu}(\vec{n}) = \widehat{T_t^*\mu}(\vec{n})$, *then* $\{\hat{T}_t: t \geqq 0\}$ *is a Bochner version of* $e^{t\hat{\mathscr{L}}}$. *Conversely, if* $\{\hat{T}_t: t \geqq 0\}$ *is a Bochner version of* $e^{t\hat{\mathscr{L}}}$ *and* $\{T_t: t \geqq 0\}$ *is defined on* $C(T)$ *by* $T_t^*\mu(\vec{n}) = \hat{T}_t\hat{\mu}(\vec{n})$, *then* $\{T_t: t \geqq 0\}$ *is a Feller version of* $e^{t\mathscr{L}}$.

PROOF. Let $\{T_t: t \geqq 0\}$ be a Feller version of $e^{t\mathscr{L}}$ and define $\hat{T}_t\hat{\mu}(\vec{n}) = \widehat{T_t^*\mu}(\vec{n})$. It is obvious that $\{\hat{T}_t: t \geqq 0\}$ is a semigroup of contractions on $\hat{M}(T^d)$ and that $\widehat{M^+}(T^d)$ is invariant under $\{\hat{T}_t: t \geqq 0\}$. Moreover, if $\{\hat{\mu}_m\}_1^\infty$ is a bounded sequence in $\hat{M}(T)$ such that $\hat{\mu}_m(\vec{n}) \to \hat{\mu}(\vec{n})$ for all $\vec{n} \in Z^d$, then $\mu_m \to \mu$ in the weak* topology on $M(T)$ and so

$$\begin{aligned}\hat{T}_t\hat{\mu}_m(\vec{n}) &= \int e_{\vec{n}}(\vec{\theta})T_t^*\mu_m\,(d\theta) = \int T_te_{\vec{n}}(\vec{\theta})\mu_m(d\theta) \to \int T_te_{\vec{n}}(\theta)\mu\,(d\vec{\theta})\\ &= \int e_{\vec{n}}(\vec{\theta})T_t^*\mu\,(d\vec{\theta}) = \hat{T}_t\hat{\mu}(\vec{n})\end{aligned}$$

since $T_te_{\vec{n}} \in C(T^d)$. Finally, note that $\hat{T}_te_{\vec{\theta}}(\vec{n}) = \widehat{T_t^*\delta_{\vec{\theta}}}(\vec{n}) = T_te_{\vec{n}}(\vec{\theta})$ and so

$$\begin{aligned}\hat{T}_te_{\vec{\theta}}(\vec{n}) - e_{\vec{\theta}}\,(\vec{n}) &= \int_0^t T_s\mathscr{L}e_{\vec{n}}(\vec{\theta})\,ds\\ &= |\vec{n}|^2\int_0^t \sum_{\vec{k}} p_{\vec{k}}T_s(e_{\vec{n}+\vec{k}} - e_{\vec{n}})(\vec{\theta})\,ds\\ &= |\vec{n}|^2 \sum_{\vec{k}} p_{\vec{k}}\int_0^t(\hat{T}_se_{\vec{\theta}}(\vec{n}+\vec{k}) - \hat{T}_se_{\vec{\theta}}(\vec{n}))\,ds\\ &= \int_0^t \hat{\mathscr{L}}\hat{T}_se_{\vec{\theta}}(\vec{n})\,ds.\end{aligned}$$

From this it is clear that

$$\hat{T}_t\hat{\mu}(\vec{n}) - \hat{\mu}(\vec{n}) = \int_0^t \hat{\mathscr{L}}\hat{T}_s\hat{\mu}(\vec{n})\,ds$$

if μ is a purely atomic measure with only a finite number of atoms. Since every $\mu \in M(T^d)$ is the weak* limit of such measures, it follows that (11) holds for all $\hat{\mu} \in \hat{M}(T^d)$ because $\hat{T}_t\hat{\mu}_m(\vec{n}) \to \hat{T}_t\hat{\mu}(\vec{n})$, $\hat{\mathscr{L}}\hat{T}_s\hat{\mu}_m(\vec{n}) \to \hat{\mathscr{L}}\hat{T}_s\hat{\mu}(\vec{n})$, and $\sup_{m;\, s\in[0,t]}|\hat{\mathscr{L}}\hat{T}_s\hat{\mu}_m(\vec{n})| < \infty$ if $\mu_m \to \mu$ in the weak* topology.

Next suppose $\{\hat{T}_t: t \geqq 0\}$ is a Bochner version of $e^{t\hat{\mathscr{L}}}$ and define $\widehat{T_t^*\mu}(\vec{n}) = \hat{T}_t\hat{\mu}(\vec{n})$ for $\mu \in M(T^d)$. Clearly $\{T_t^*: t \geqq 0\}$ is a nonnegative contraction semigroup on $M(T^d)$. Moreover, if $T_tf(\vec{\theta}) = \int f(\vec{\xi})T_t^*\delta_{\vec{\theta}}(d\vec{\xi})$, then it is obvious that $\{T_t: t \geqq 0\}$ is a nonnegative contraction semigroup on $C(T^d)$, since if $\vec{\theta}_m \to \vec{\theta}$, then $\widehat{T_t^*\delta_{\theta_m}}(\vec{n}) = \hat{T}_t\hat{\delta}_{\theta_m}(\vec{n}) \to \hat{T}_t\hat{\delta}_\theta(\vec{n}) = \widehat{T_t^*\delta_{\vec{\theta}}}(\vec{n})$ and so $T_t^*\delta_{\theta_m} \to T_t^*\delta_\theta$ in the weak* topology. Finally,

$$T_te_{\vec{n}}(\vec{\theta}) = \int e_{\vec{n}}(\vec{\xi})T_t^*\delta_{\vec{\theta}}(d\vec{\xi}) = \widehat{T_t^*\delta_{\vec{\theta}}}(\vec{n}) = \hat{T}_t\hat{\delta}_\theta(\vec{n}) = \hat{T}_te_{\vec{\theta}}(\vec{n})$$

and so

$$\begin{aligned}T_te_{\vec{n}}(\vec{\theta}) - e_{\vec{n}}(\vec{\theta}) = \hat{T}_te_{\vec{\theta}}(\vec{n}) - e_{\vec{\theta}}(\vec{n}) &= \int_0^t \hat{\mathscr{L}}\hat{T}_se_{\vec{\theta}}(\vec{n})\,ds\\ &= |\vec{n}|^2\int_0^t \sum_{\vec{k}} p_{\vec{k}}(\hat{T}_se_{\vec{\theta}}(\vec{n}+\vec{k}) - \hat{T}_se_{\vec{\theta}}(\vec{n}))\,ds\\ &= |\vec{n}|^2\int_0^t \sum_{\vec{k}} p_{\vec{k}}T_s(e_{\vec{n}+\vec{k}} - e_{\vec{n}})(\vec{\theta})\,ds\\ &= \int_0^t T_s\mathscr{L}e_{\vec{n}}(\vec{\theta})\,ds.\end{aligned}$$

It is easy to pass from this to the statement that $T_t f - f = \int_0^t T_s \mathscr{L} f\, ds$ for all $f \in C^2(T^d)$. Q.E.D.

The rest of this section is devoted to examining in detail the duality just established when $d = 1$. Let $\{p_k : k \in Z\}$ be a symmetric, aperiodic probability measure on Z, set $\varphi(\theta) = \sum_k p_k e_n(\theta)$, and define $\mathscr{L} = (1 - \varphi(\theta))d^2/d\theta^2$ on $C^2(T)$. Since nothing of interest comes out of the case when there is exactly one Feller version of $e^{t\mathscr{L}}$, we will assume that

$$\int \frac{1}{1 - \varphi(\theta)}\, d\theta < \infty \quad \text{and} \quad 1 - \varphi(\theta) > 0 \text{ for } \theta \neq 0. \tag{12}$$

The next result characterizes all the Feller versions of $e^{t\mathscr{L}}$ when (12) holds; it is an extension of the reasoning given in Girsanov **[2]**.

LEMMA. *If $\{T_t : t \geqq 0\}$ is a Feller version of $e^{t\mathscr{L}}$, then the generator of $\{T_t : t \geqq 0\}$ is a member of the family $\{A^{(\alpha)} : 0 \leqq \alpha \leqq \infty\}$ where $A^{(\alpha)} = (d/dm_\alpha)d/d\theta$ and m_α is given by*

$$m_\alpha(d\theta) = \frac{d\theta}{1 - \varphi(\theta)} + \frac{1}{\alpha}\delta_0(d\theta) \tag{13}$$

and the domain $\mathscr{D}(A^{(\alpha)})$ of $A^{(\alpha)}$ is the set of $f \in C^2(T\backslash\{0\}) \cap C(T)$ such that $(1 - \varphi(\cdot))f''(\cdot)$ admits a continuous extension to T and

$$\lim_{\theta \to 0;\, \theta \neq 0} (1 - \varphi(\theta))f''(\theta) = \lim_{\varepsilon \searrow 0} \frac{f'(\varepsilon) - f'(-\varepsilon)}{2\int_0^\varepsilon d\theta/(1 - \varphi(\theta)) + 1/\alpha}. \tag{14}$$

(If $\alpha = 0$, (14) means that $\lim_{\theta \to 0;\theta \neq 0}(1 - \varphi(\theta))f''(\theta) = 0$. If $\alpha = \infty$, (14) implies that $f \in C^2(T\backslash\{0\}) \cap C^1(T)$.)

PROOF. It is easy to see that $\{T_t : t \geqq 0\}$ is the semigroup of a continuous Markov process $\theta(\cdot)$ on T. There are two possibilities. Either 0 is an absorbing point of $\theta(\cdot)$ or it is not. If it is, then one can easily check that $A^{(0)}$ is the generator of $\{T_t : t \geqq 0\}$. If 0 is not absorbing, then by choosing $f \in C^\infty(T)$ so that $f'(\theta) = 1$ near zero, we see that $Af = \mathscr{L}f = 0$ near zero; and therefore it follows that all points of T are regular for $\theta(\cdot)$ and $p(d\theta) = d\theta$ is the scale measure. Finally, an obvious argument shows that the speed measure $m(d\theta)$ is $d\theta/(1 - \varphi(\theta))$ away from 0, and therefore must be of the form given in (13) for some $\alpha > 0$. Q.E.D.

It is easy to construct the process generated by $A^{(\alpha)}$ in terms of the standard Brownian motion $\beta(\cdot)$ on T. Indeed, let $l_\theta(\cdot)$ be the local time functional of $\beta(\cdot)$ at $\theta \in T$ and let $\tau^{(\alpha)}(\cdot)$ be the inverse of $\frac{1}{2}\int l_\theta(\cdot) m_\alpha(d\theta)$. Then $\theta^{(\alpha)}(\cdot) = \beta \circ \tau^{(\alpha)}(\cdot)$ is the process going with $A^{(\alpha)}$. From this representation, we can conclude that all the $\theta^{(\alpha)}(\cdot)$ are identical until they reach 0, $\theta^{(0)}(\cdot)$ is absorbed at 0, $\theta^{(\infty)}(\cdot)$ passes right through zero and does not spend a positive time there, and $\theta^{(\alpha)}(\cdot)$, $0 < \alpha < \infty$, "sticks" at 0 in the sense that it spends a Cantor set of positive measure there.

Set $T_t^{(\alpha)} = e^{tA^{(\alpha)}}$, $t \geqq 0$. We are now going to describe the dual semigroups $\{\hat{T}_t^{(\alpha)} : t \geqq 0\}$, $0 \leqq \alpha \leqq \infty$. First, we will identify $\{\hat{T}_t^{(\infty)} : t \geqq 0\}$. Define $\{\hat{P}_n^{(N)} : n \in Z\}$ and $\zeta^{(N)}$, $N \geqq 1$, as before. Then it is easy to check that if

$$\hat{S}_t \hat{\mu}(n) \equiv \lim_{N \nearrow \infty} \hat{E}_n^{(N)}[\hat{\mu}(n(t)), \zeta^{(N)} > t]$$

for bounded $\hat{\mu}: Z \to \mathscr{R}$, then $\{\hat{S}_t : t \geqq 0\}$ is a Markov (nonconservative) semigroup on $B(Z)$ (the bounded function on Z). (In fact, $\{\hat{S}_t : t \geqq 0\}$ is the semigroup associated with the minimal process having generator $\hat{\mathscr{L}}$.) To see that $\hat{S} = \hat{T}^{(\infty)}$, it is certainly enough to check that

$$\int_0^\infty e^{-\lambda t}\, T_t^{(\infty)} e_n(\theta)\, dt = \int_0^\infty e^{-\lambda t}\, \hat{S}_t e_\theta(n)\, dt \quad \text{for all } \lambda > 0.$$

But if $u_\lambda(\theta, n) = \int_0^\infty e^{-\lambda t}\, T_t^{(\infty)} e_n(\theta)\, dt$, then, from (5), it is obvious that

$$u_\lambda(\theta, n) = \hat{E}_n^{(N)}\left[\int_0^{\zeta^{(N)}} e^{-\lambda t} e_\theta(n(t))\, dt\right] + \hat{E}_n^{(N)}[e^{-\lambda\zeta^{(N)}} u_\lambda(\theta, n(\zeta^{(N)}))].$$

Clearly $\hat{E}_n^{(N)}[\int_0^{\zeta^{(N)}} e^{-\lambda t} e_\theta(n(t))\, dt] \to \int_0^\infty e^{-\lambda t} S_t e_\theta(n)\, dt$ as $N \to \infty$. Moreover, from the fact that the transition function $P^{(\infty)}(t, \theta, \cdot)$ of $T_t^{(\infty)}$ can have no mass at 0 for $t > 0$, one sees that $P^{(\infty)}(t, \theta, \cdot)$ has an $L^1(T)$-density.[2] Hence, by the Riemann-Lebesgue lemma, $T_t^{(\infty)} e_n(\theta)$ tends to 0 as $|n| \to \infty$, and therefore $u_\lambda(\theta, n) \to 0$ as $|n| \to \infty$. Thus $E_n^{(N)}[e^{-\lambda\zeta^{(N)}} u_\lambda(\theta, n(\zeta^{(N)}))] \to 0$ as $N \to \infty$, and the identification is complete.

Next, let $0 \leqq \alpha < \infty$ and set $R_\lambda^{(\alpha)} = \int_0^\infty e^{-t\lambda} T_t^{(\alpha)}\, dt$ for $\lambda > 0$. Then, by the strong Markov property,

$$R_\lambda^{(\alpha)} e_m(\theta) = R_\lambda^{(\infty)} e_m(\theta) + E_\theta^{(\alpha)}[e^{-\lambda\tau}](R_\lambda^{(\alpha)} e_m(0) - R_\lambda^{(\infty)} e_m(0)) \tag{15}$$

where $\tau = \inf\{t \geqq 0: \theta(t) = 0\}$. Note that $E_\theta^{(\alpha)}[e^{-\lambda\tau}]$ is independent of $0 \leqq \alpha \leqq \infty$; and, in fact, $E_\theta^{(\alpha)}[e^{-\lambda\tau}]$ can easily be identified with the unique $h_\lambda \in C^2(T\backslash\{0\}) \cap C(T)$ such that

$$(\lambda - \mathscr{L})h_\lambda(\theta) = 0, \qquad \theta \in T\backslash\{0\}, \quad \text{and} \quad \lim_{\theta \to 0} h_\lambda(\theta) = 1. \tag{16}$$

Thus, if $\hat{R}_\lambda^{(\alpha)}(m, n) = \widehat{R_\lambda^{(\alpha)} e_m}(n)/2\pi$, $0 \leqq \alpha \leqq \infty$ and $m, n \in Z$, then

$$\hat{R}_\lambda^{(\alpha)} \hat{\mu}(m) = \sum_{n \in Z} \hat{R}_\lambda^{(\alpha)}(m, n)\hat{\mu}(n), \qquad \hat{\mu} \in \hat{\mathscr{M}}(T), \tag{17}$$

and

$$\hat{R}_\lambda^{(\alpha)}(m, n) = \hat{R}_\lambda^{(\infty)}(m, n) + \frac{(R_\lambda^{(\alpha)} e_m(0) - R_\lambda^{(\infty)} e_m(0))}{2\pi} \hat{h}_\lambda(n). \tag{18}$$

We have already seen that $\hat{R}_\lambda^{(\infty)}$ is the resolvent for the minimal process on Z generated by $\hat{\mathscr{L}}$. Thus it only remains to compute $R_\lambda^{(\alpha)} e_m(0) - R_\lambda^{(\infty)} e_m(0)$ and $\hat{h}_\lambda(n)$.

LEMMA. *Let* $x_\lambda(n) = 1 - \lambda(\hat{R}_\lambda^{(\infty)} 1)(n)$, $n \in Z$, *and put*

$$a_\lambda^{-1} = \frac{1}{\lambda} \sum_n p_n x_\lambda(n) + \sum_{n \neq 0} \frac{x_\lambda(n)}{n^2}.$$

Then

$$\begin{aligned} \hat{h}_\lambda(n) &= 2\pi a_\lambda \frac{x_\lambda(n)}{n^2} && \textit{if } n \neq 0, \\ &= \frac{2\pi a_\lambda}{\lambda} \sum_k p_k x_\lambda(k) && \textit{if } n = 0. \end{aligned} \tag{19}$$

[2]One must use here the fact that $1 - \varphi(\theta) > 0$ for $\theta \neq 0$ to show that $P^{(\infty)}(t, \theta, \cdot) \ll d\theta$ away from 0.

In particular, $\hat{h}_\lambda(n) > 0$ *for all* n *and*

(20) $$\sum_n \hat{h}_\lambda(n) = 2\pi.$$

PROOF. Note that $h''_\lambda = \lambda h_\lambda/(1-\varphi) \in L^1(T)$. Thus $\widehat{h''_\lambda}(n) \to 0$ as $|n| \to \infty$. Moreover, since $h_\lambda \in C(T)$, if $n \neq 0$, then

$$\widehat{h_\lambda}(n) = -\frac{1}{in}\int_0^{2\pi} h'_\lambda(\theta)e_n(\theta)\,d\theta = \frac{1}{n^2}\left(h'_\lambda(2\pi-) - h'_\lambda(0+) - \widehat{h''_\lambda}(n)\right)$$

and

$$h'_\lambda(2\pi-) - h'_\lambda(0+) = \int_0^{2\pi} h''_\lambda(\theta)\,d\theta = \widehat{h''_\lambda}(0).$$

Thus

(21) $$\widehat{h_\lambda}(n) = \frac{\widehat{h''_\lambda}(0) - \widehat{h''_\lambda}(n)}{h^2}, \qquad n \neq 0.$$

Next observe that if $f \in L^2(T)$, then $\widehat{(1-\varphi)f}(n) = \sum_k p_k(\hat{f}(n) - \hat{f}(n+k))$. Since $\{p_k: k \in Z\} \in l^1(Z)$, it follows that this equation also holds for $f \in L^1(T)$. Thus

$$\widehat{(1-\varphi)h''_\lambda}(n) = \sum_k p_k(\widehat{h''_\lambda}(n) - \widehat{h''_\lambda}(n+k)).$$

But $(1-\varphi)h''_\lambda = \lambda h_\lambda$ on $T\backslash\{0\}$, and therefore

(22) $$\widehat{\lambda h_\lambda}(n) = \sum_k p_k(\widehat{h''_\lambda}(n) - \widehat{h''_\lambda}(n+k)), \qquad n \in Z.$$

Combining (21) and (22), we see that, for $n \neq 0$, $\lambda(\widehat{h''_\lambda}(0) - \widehat{h''_\lambda}(n)) = -\hat{\mathscr{L}}\widehat{h''_\lambda}(n)$. Since this equation is trivially true for $n = 0$, we conclude that

(23) $$(\lambda - \hat{\mathscr{L}})\widehat{h''_\lambda}(n) = \lambda\widehat{h''_\lambda}(0), \qquad n \in Z.$$

Combining (23) with the fact that $\widehat{h''_\lambda}(n) \to 0$ as $|n| \to \infty$, one can easily show that $\widehat{h''_\lambda}(n) = \lambda\widehat{h''_\lambda}(0)\,(\widehat{R_\lambda^{(\infty)}}1)(n)$, $n \in Z$, and therefore

(24) $$\widehat{h_\lambda}(n) = \frac{\widehat{h''_\lambda}(0) - \widehat{h''_\lambda}(n)}{n^2} = \widehat{h''_\lambda}(0)\,\frac{x_\lambda(n)}{n^2}, \qquad n \neq 0.$$

Taking $n = 0$ in (22), we also have

(25) $$\hat{h}_\lambda(0) = \widehat{h''_\lambda}(0)\sum_k p_k x_\lambda(k).$$

Since $h_\lambda(0) = 1$ and, as we have just explicitly shown, $\{\hat{h}_\lambda(n): n \in Z\} \in l^1(Z)$, (20) is obvious. In other words,

(26) $$\widehat{h''_\lambda}(0)\left[\frac{1}{\lambda}\sum_k p_k x_\lambda(k) + \sum_{k\neq 0}\frac{x_\lambda(k)}{k^2}\right] = 2\pi,$$

and this clearly completes the proof. Q.E.D.

THEOREM. *Let* $\hat{R}_\lambda^{(\alpha)}$, $\lambda > 0$ *and* $0 \leqq \alpha \leqq \infty$, *be the resolvent of* $\{\hat{T}_t^{(\alpha)}: t \geqq 0\}$. *Then* $\hat{R}_\lambda^{(\infty)}$ *is the resolvent of the minimal process on* Z *generated by* $\hat{\mathscr{L}}$ *(i.e., the one which is killed as soon as it explodes). Moreover, if* $\hat{R}_\lambda^{(\alpha)}(m, n) \equiv \widehat{R_\lambda^{(\alpha)}e_m}(n)/2\pi$, *then* (17) *holds and*

$$\hat{R}_\lambda^{(\alpha)}(m, n) = \widehat{R^\infty}(m, n) + \frac{x_\lambda(m)\eta_\lambda(n)}{\lambda \sum_k \eta_\lambda(k) + 2\pi\alpha}, \tag{27}$$

where $x_\lambda(n) \equiv 1 - \lambda(\widehat{R_\lambda^{(\infty)}1})(n)$ *and*

$$\eta_\lambda(n) \equiv \frac{1}{\lambda}\sum_k p_k x_\lambda(k) \quad \text{if } n = 0,$$
$$\equiv x_\lambda(n)/n^2 \qquad \text{if } n \neq 0.$$

PROOF. Starting from (15), we have

$$(R_\lambda^{(\alpha)}e_m)'(0+) - (R_\lambda^{(\alpha)}e_m)'(2\pi-) = h_\lambda'(0+) - h_\lambda'(2\pi-)(R_\lambda^{(\alpha)}e_m(0) - R_\lambda^{(\infty)}e_m(0)),$$

since $R_\lambda^{(\infty)}e_m \in C^1(T)$. Since $h_\lambda'(0+) - h_\lambda'(2\pi-) = -\widehat{h_\lambda''}(0)$,

$$\alpha[(R_\lambda^{(\alpha)}e_m)'(0+) - (R_\lambda^{(\alpha)}e_m)'(2\pi-)] = A^{(\alpha)}R_\lambda^{(\alpha)}e_m(0),$$

and $(\lambda - A^{(\alpha)})P_\lambda^{(\alpha)}e_m(0) = 1$, one has

$$\widehat{R_\lambda^{(\alpha)}}e_m(0) - \widehat{R_\lambda^{(\infty)}}e_m(0) = x_\lambda(m)/(\lambda + h_\lambda''(0)\alpha).$$

From the lemma above, we also have $\hat{h}_\lambda(n) = 2\pi\eta_\lambda(n)/\sum_k \eta_\lambda(k)$. Combining these with (18), we arrive at

$$\widehat{R_\lambda^{(\alpha)}}(m, n) = \widehat{R_\lambda^{(\infty)}}(m, n) + \frac{x_\lambda(m)\eta_\lambda(n)}{(\lambda + h_\lambda''(0)\alpha)\sum_k \eta_\lambda(k)}.$$

The theorem is now completed by an application of (26). Q.E.D.

REMARK (1). It may, at first, seem odd that $\{\hat{T}_t^{(\alpha)}: t \geqq 0\}$ is (sub-) Markovian as well as being a Bochner semigroup. However, this fact is less surprising when one notes that $\{T_t^{(\infty)}: t \geqq 0\}$ is the weak limit as $\varepsilon \searrow 0$ of $\{e^{t\mathscr{L}_\varepsilon}: t \geqq 0\}$, where $\mathscr{L}_\varepsilon = (1 + \varepsilon - \varphi)\,\partial^2/\partial\theta^2$, and $\{T_t^{(0)}: t \geqq 0\}$ is the weak limit as $r \nearrow 1$ of $\{e^{t\mathscr{L}_{(r)}}: t \geqq 0\}$, where $\mathscr{L}_{(r)} = (1 - p_r\varphi)\,\partial^2/\partial\theta^2$ and $p_r(\theta)$ is the Poisson kernel. (The verification of these assertions is easily accomplished by writing the generators involved in Feller form and observing that the associated speed and scale measure converge (weakly).) From this it follows easily that the corresponding dual semigroups converge, and therefore the limiting dual semigroups are nonnegativity preserving since the approximating ones are.

REMARK (2). Using our duality results, it is possible to show that, for each $0 \leqq \alpha \leqq \infty$, $\{T_t^{(\alpha)}: t \geqq 0\}$ is ergodic and that the unique invariant measure for $\{T_t^{(\alpha)}: t \geqq 0\}$ is $\mu_\alpha(d\theta) = c_\alpha m_\alpha(d\theta)$, where $c_\alpha = c_\infty\alpha/c_\infty + \alpha$ and $c_\infty^{-1} = \int d\theta/(1 - \varphi(\theta))$. (That μ_α is indeed invariant under $\{T_t^{(\alpha)}: t \geqq 0\}$ is obvious from the fact that $\int A^{(\alpha)}f(\theta)\mu_\alpha(d\theta) = 0$ for all $f \in \mathscr{D}(A^{(\alpha)})$. Thus the real question is whether μ_α is the *only* invariant measure.) Clearly, since $\theta^{(0)}(\cdot)$ is absorbed at 0 and gets to 0 with probability 1, $T_t^{(0)}f \to f(0) = \int f(\theta)\mu_0(d\theta)$ as $t \to \infty$ for all $f \in C(T)$. Thus the case when $\alpha = 0$ is trivial. When $\alpha = \infty$, we have for any probability measure μ on T

$$\widehat{T_t^{(\infty)*}\mu}(n) = \lim_{N \nearrow \infty} \hat{E}_n^{(N)}[\hat{\mu}(n(t)), \zeta^{(N)} > t].$$

Observe that

$$\hat{E}_n^{(N)}\left[\int_0^\infty \mathscr{X}_{(0,N)}(|n(s)|)\,ds, \zeta^{(N)} > t\right] \leqq 2\left(\sum_1^\infty \frac{1}{k^2}\right)\int \frac{1}{1-\varphi(\theta)}\,d\theta,$$

and therefore

$$\limsup_{t\nearrow\infty} \sup_N \sup_n \hat{P}_n^{(N)}(n(t) \neq 0, \zeta^{(N)} > t) = 0.$$

Thus

$$\lim_{t\nearrow\infty} \widehat{T_t^{(\infty)*}\mu}(n) = \lim_{N\nearrow\infty} \hat{P}_n^{(N)}(\hat{\tau}_0 < \infty),$$

where $\hat{\tau}_0 = \inf\{t \geqq 0: n(t) = 0\}$. But $\lim_{N\nearrow\infty} \hat{P}_n^{(N)}(\hat{\tau}_0 < \infty)$ is the same as the probability that the random walk determined by $\{p_k : k \in Z\}$ ever reaches 0 starting from n, and so

$$\lim_{t\nearrow\infty} \widehat{T_t^{(\infty)*}\mu}(n) = \frac{(1/2\pi)\int e^{in\theta}\,d\theta/(1-\varphi(\theta))}{(1/2\pi)\int d\theta/(1-\varphi(\theta))} = \hat{\mu}_\infty(n).$$

This shows that $T_t^{(\infty)}f \to \int f\mu_\infty(d\theta)$ for $f \in C(T)$.

To handle $0 < \alpha < \infty$, observe that from (27)

$$R_\lambda^{(\alpha)} = (1-\gamma_\lambda(\alpha))R_\lambda^{(\infty)} + \gamma_\lambda(\alpha)R_\lambda^{(0)},$$

where $\gamma_\lambda(\alpha) = \lambda\beta_\lambda/(\lambda\beta_\lambda + 2\pi\alpha)$ and $\beta_\lambda = \lambda^{-1}\sum_k p_k x_\lambda(k) + \sum_{k\neq 0}(x_\lambda(k)/k^2)$. From the preceding, we know that $\lambda R_\lambda^{(\infty)} f \to \int f\mu_\infty(d\theta)$ and $\lambda R_\lambda^{(0)} f \to \int f\mu_0(d\theta)$ as $\lambda \searrow 0$ for all $f \in C(T)$. Thus we will have shown that $\lambda R_\lambda^{(\alpha)} f \to \int f\mu_\alpha(d\theta)$ as $\lambda \searrow 0$ once we prove that $\lim_{\lambda\searrow 0}\gamma_\lambda(\alpha) = c_\infty/(c_\infty + \alpha)$. But

$$\begin{aligned}\lim_{\lambda\searrow 0} \lambda\beta_\lambda &= \sum_0^\infty p_k \lim_{\lambda\searrow 0} x_\lambda(k) = 1 - \sum_0^\infty p_k \lim_{\lambda\searrow 0} \lambda\widehat{R_\lambda^{(\infty)}e_k}(0)\\ &= 1 - \sum_0^\infty p_k\hat{\mu}_\infty(k) = 1 - c_\infty \widehat{\frac{\varphi}{1-\varphi}}(0) = c_\infty \int_0^{2\pi} \frac{1-\varphi}{1-\varphi}\,d\theta = 2\pi c_\infty.\end{aligned}$$

Note that the ergodicity for the cases when $\alpha = 0$ or ∞ seems to be "better" than when $0 < \alpha < \infty$ in the sense that in those two cases we have the convergence of the semigroup itself to the invariant measure, whereas in the other cases we have only shown that $(1/t)\int_0^t T_s^{(\alpha)}\,ds \to \mu_\alpha$. We have not been able to find out whether this apparent difference is really there or is only an artefact of our method.

REMARK (3). The possible modes of behavior of $\mathscr{L}$-diffusions at 0 correspond to modes of behavior of $\hat{\mathscr{L}}$-chains at ∞. Let us examine (briefly and without proofs) how things look from the point of view of the boundary theory of chains. Aficionados of chain theory will of course have met formula (27) on many occasions—for example, in Reuter [**10**].

Let $n^{(0)}(t)$ be a chain with minimal state-space Z and with resolvent $\hat{R}^{(0)}(\cdot,\cdot)$. Then the point ∞ in the one-point compactification of Z is a *regular* boundary point for $n^{(0)}(\cdot)$ and we may introduce the local time $\Lambda_\infty^{(0)}(\cdot)$ at ∞ for $n^{(0)}(\cdot)$. We assume $\Lambda_\infty^{(0)}(\cdot)$ normalised in the usual "chain" fashion so that, for example,

$$\Lambda_\infty^{(0)}(t) = \lim_{\lambda\nearrow\infty} \lambda\int_0^t x_\lambda(n^{(0)}(s))\,ds,$$

the limit existing uniformly in probability on compact intervals. Let ξ be an exponentially distributed variable of rate $2\pi\alpha$ and independent of $n^{(0)}(\cdot)$. Then we get a chain $n^{(\alpha)}(\cdot)$ with resolvent $\hat{R}^{(\alpha)}(\cdot, \cdot)$ by killing $n^{(0)}(\cdot)$ at the instant when the local time $\Lambda_\infty^{(0)}$ first reaches the value ξ.

The boundary point ∞ of $n^{(0)}(\cdot)$ may be regarded as a *reflecting* barrier. This idea can be made precise as follows: If we restrict attention to excursions of $n^{(0)}(\cdot)$ from ∞ which do not become absorbed at 0, then the Itô excursion law **[9]** of $n^{(0)}(\cdot)$ at ∞ is invariant under time-reversal. The time-reversibility is implicit in the symmetrisability condition

$$m^{-2}\hat{R}_\lambda^{(0)}(m, n) = n^{-2}\hat{R}_\lambda^{(0)}(n, m) \qquad (m, n \in Z\backslash\{0\}).$$

Because $n^{(\alpha)}$ is $n^{(0)}$ killed at ∞ at rate $2\pi\alpha$, the point ∞ acts as an *elastic* barrier for $n^{(\alpha)}$. By utilising Feller's idea **[8]** of normal derivative D at ∞, we can obtain the infinitesimal generator of $n^{(\alpha)}$ by restricting $\hat{\mathscr{L}}$ via an elastic barrier condition

$$D\hat{f}(\infty) + 2\pi\alpha\hat{f}(\infty) = 0,$$

which is clearly the appropriate "dual" form of (14).

Finally, it should be noted that the duality between $\theta^{(\alpha)}$ and $n^{(\alpha)}$ hinges on a striking duality between the Markov random sets $\{t: \theta^{(\infty)}(t) = 0\}$ and $\{t: n^{(0)}(t) = \infty\}$ which is easily described in terms of the associated Lévy kernels.

2. Dual processes and ergodic properties. In this section we give several examples of how dual processes can be used to determine the ergodic properties of the diffusion.

Consider the generator $\mathscr{L} = (\gamma - d^{-1}\sum_1^d \cos\theta_i)\Delta$ on T^d. In view of the preceding considerations, we know that the unique Markov version $\{T_t: t \geqq 0\}$ of $e^{t\mathscr{L}}$ has the property that

$$\widehat{T_t^*\mu}(\vec{n}) = \hat{E}_{\vec{n}}\left[\hat{\mu}(\vec{n}(t))\exp\left(-(\gamma - 1)\int_0^t |\vec{n}(s)|^2\,ds\right)\right], \tag{28}$$

where $\hat{E}_{\vec{n}}[\ \ ]$ is expectation with respect to the distribution $P_{\vec{n}}$ of the process obtained by speeding up the standard symmetric random walk on Z^d by $|\vec{n}|^2$ at $\vec{n} \neq 0$ and absorbing it at 0 (note that (10) is satisfied in this case and so we can let $N \to \infty$ in (6)). If $\gamma > 1$, then as $t \nearrow \infty$ we get

$$\lim_{t\nearrow\infty} \widehat{T_t^*\mu}(\vec{n}) = \hat{E}_{\vec{n}}\left[\exp\left(-(\gamma - 1)\int_0^{\tau_0} |\vec{n}(s)|^2\,ds\right)\right], \tag{29}$$

where $\tau_0 = \inf\{t \geqq 0: \vec{n}(t) = 0\}$. Thus when $\gamma > 1$, we can conclude from (28) that $\{T_t: t \geqq 0\}$ is ergodic. Of course the same conclusion can be reached in this case by more powerful analytic considerations. To take full advantage of the dual process technique, set $\gamma = 1$. Then (28) becomes

$$\widehat{T_t^*\mu}(\vec{n}) = \hat{E}_{\vec{n}}[\hat{\mu}(\vec{n}(t))]. \tag{30}$$

When $d \leqq 2$, the standard symmetric random walk is recurrent, and so $\hat{P}_{\vec{n}}(\vec{n}(t) = 0) \to 1$ as $t \nearrow \infty$ for all $\vec{n} \in Z^d$. Thus when $d \leqq 2$, (30) shows that $\widehat{T_t^*\mu}(\infty) \to 1$ as $t \nearrow \infty$, and so $T_t\mu$ tends to the δ-mass at 0. When $d \geqq 3$, the symmetric random walk is no longer recurrent. Nonetheless, (30) still yields information. To be specific, take δ_0 equal to the δ-mass at 0 and let λ be the Haär measure on T^d. Then (30) shows that $T_t^*\delta_0 = \delta_0$ and $T_t^*\lambda \to \mu_1$, where $\hat{\mu}_1(\vec{n}) = \hat{P}_{\vec{n}}(\tau_0 < \infty)$. Thus δ_0 and μ_1 are distinct invariant measures for $\{T_t: t > 0\}$. To see that all other invariant measures for $\{T_t: t > 0\}$ are convex combinations of these, let μ be an invariant measure which is singular with respect to δ_0. Then $\mu = T_1^*\mu$ and it is easy to see that for $\vec{\theta} \neq 0$ the transition function for the diffusion generated by $\mathscr{L}$ is absolutely continuous with respect to λ. Thus, $\mu \ll \lambda$ and by the Riemann-Lebesgue lemma, $\hat{\mu}(\vec{n}) \to 0$ as $|\vec{n}| \to \infty$. From this and (30) it is easy to see that

$$\hat{\mu}(\vec{n}) = \widehat{T_t^*\mu}(\vec{n}) = \hat{P}_{\vec{n}}(\tau_0 \leqq t) + \hat{E}_{\vec{n}}[\hat{\mu}(\vec{n}(t)), \tau_0 > t] \to \hat{P}_{\vec{n}}(\tau_0 < \infty) = \hat{\mu}_1(\vec{n}),$$

since $|\vec{n}(t)| \to \infty$ on $\{\tau_0 = \infty\}$. Finally, since $(1 - (1/d)\sum_1^d \cos\theta_i)^{-1}\, d\vec{\theta}$, properly normalised, is an invariant measure for $\{T_t: t \geqq 0\}$ which is absolutely continuous with respect to λ, we can identify it as μ_1.

Next consider the operator $\mathscr{L} = (1 - \cos\theta)\,\partial^2/\partial\theta^2 + \lambda \sin\theta\, \partial/\partial\theta$ on T^1, where $|\lambda| \leqq 1$, and let $\{T_t: t > 0\}$ be the unique Markov semigroup generated by $\mathscr{L}$. By exactly the same procedure as we used in §1, one can show that T_t^* satisfies

$$\widehat{T_t^*\mu}(n) = \hat{E}_n[\hat{\mu}(n(t))], \tag{31}$$

where $\hat{E}_n[\cdot]$ stands for expectation with respect to $\hat{P}_n$ and $\{\hat{P}_n: n \in Z\}$ is the Markov family generated by $\hat{\mathscr{L}}$, where

$$\hat{\mathscr{L}}f(n) = n^2\left(\frac{(1 + \lambda/n)f(n+1) + (1 - \lambda/n)f(n-1)}{2} - f(n)\right) = n^2\mathscr{A}f(n). \tag{32}$$

The process generated by $\mathscr{A}$ is a simple birth and death process, and so it is easy to see that $\hat{P}_n(\tau_0 < \infty) < 1$ for $n \neq 0$ if and only if $\lambda \geqq \frac{1}{2}$; and $\hat{P}_n(\tau_0 = \infty$ and $|n(t)| \nrightarrow \infty) = 0$. Thus, from (31), we see that $T_t^*\mu \to \delta_0$ as $t \nearrow \infty$ if $\lambda < \frac{1}{2}$; and, proceeding as in the preceding example, $\widehat{T_t^*\mu}(n) \to \alpha + (1 - \alpha)\hat{P}_n(\tau_0 < \infty)$ when $\lambda \geqq \frac{1}{2}$, where $\mu = \alpha\delta_0 + (1 - \alpha)\mu'$ is the Lebesgue decomposition of μ with respect to δ_0. This example can also be handled by writing $\mathscr{L}$ in Feller form. An example of the same sort which is not amenable to "Fellerization" is the operator $\mathscr{L} = (1 - \cos\theta_2)\,\partial^2/\partial\theta_1^2 + (1 - \cos\theta_1)\partial^2/\partial\theta_2^2$ on T^2. Here (30) holds when $\hat{E}_n$ is associated with the $\hat{\mathscr{L}}$ given by

$$\hat{\mathscr{L}}f(\vec{n}) = n_1^2\left(\frac{f(n_1, n_2 + 1) + f(n_1, n_2 - 1)}{2} - f(n_1, n_2)\right) + n_2^2\left(\frac{f(n_1 + 1, n_2) + f(n_1 - 1, n_2)}{2} - f(n_1, n_2)\right).$$

Since the counting measure on Z^2 is an invariant measure for the process associated with $\hat{\mathscr{L}}$ and $Z^2\backslash\{0\}$ is a closed set of communicating states for this process, $\hat{P}_{\vec{n}}(0 < |\vec{n}(t)| < N) \to 0$ as $t \nearrow \infty$ for each $N \geqq 1$. Using these facts, one can now show that $T_t^*\mu \to \alpha\delta_0 + (1 - \alpha)\lambda$ as $t \nearrow \infty$, where $\mu = \alpha\delta_0 + (1 - \alpha)\mu'$ is the Lebesgue decomposition of μ with respect to δ_0. (The only pitfall here is the proof that the transition function $P(t, \vec{\theta}, \cdot)$ for the $\mathscr{L}$-process is absolutely continuous with respect to λ for $\vec{\theta} \neq 0$. But the $\mathscr{L}$-process never reaches 0 if it does not start there and $\mathscr{L}$ can be written in the form $2(X_1^2 + X_2^2)$, where $X_1 = (\sin(\theta_2/2))\partial/\partial\theta_1$ and $X_2 = (\sin(\theta_1/2))\partial/\partial\theta_2$, and therefore the results of Ichihara and Kunita **[4]** are applicable.)

Even if the operator $\mathscr{L}$ does not have a dual process in the above sense because some of the coefficients in the Fourier expansions of the drift or diffusion terms have the wrong sign, one may still be able to find a slightly modified dual process. Interestingly, the necessary modification sometimes provides a rather subtle proof of ergodicity, as the next example shows.

Let

$$\mathscr{L} = \left(1 - \frac{-\cos\theta_1 + \cos\theta_2 + \cos\theta_3}{3}\right)\frac{\partial^2}{\partial\theta_1^2} + \left(1 - \frac{\cos\theta_1 - \cos\theta_2 + \cos\theta_3}{3}\right)\frac{\partial^2}{\partial\theta_2^2} + \left(1 - \frac{\cos\theta_1 + \cos\theta_2 - \cos\theta_3}{3}\right)\frac{\partial^2}{\partial\theta_3^2}$$

on T^3. If $(\vec{n}, a) \in Z^3 \times \{-1, 1\}$ let

$$e_{\vec{\theta}}(\vec{n}, a) = E_{(\vec{n},a)}(\vec{\theta}) = ae^{i\vec{n}\cdot\vec{\theta}}.$$

Define an operator $\hat{\mathscr{L}}$ on $\mathscr{B}(Z^3 \times \{-1, 1\})$ by

$$\hat{\mathscr{L}} f(\vec{n}, a) = |\vec{n}|^2 \sum_{i=1}^{3}\sum_{j=1}^{3}\sum_{k=1}^{2} \frac{(n_i)^2}{6|\vec{n}|^2}[f(\vec{n} + (-1)^k u_i, a\alpha_{ij}) - f(\vec{n}, a)],$$

where u_j is the unit vector in the jth direction and

$$\alpha_{ij} = 1, \qquad i \neq j,$$
$$= -1, \qquad i = j.$$

It is easily checked that $\mathscr{L}e_{(\vec{n},a)}(\vec{\theta}) = \hat{\mathscr{L}}e_{\vec{\theta}}(\vec{n}, a)$. Moreover $\hat{\mathscr{L}}$ generates a semigroup $\hat{T}_t$ of a Markov process $(\vec{n}(t), a(t))$ on $Z^3 \times \{-1, 1\}$ and just as in (30) we have

$$a\,\widehat{T_t^*\mu}(n) = \hat{E}_{(\vec{n},a)}[a(t)\hat{\mu}(\vec{n}(t))]. \tag{33}$$

From the definition of $\hat{\mathscr{L}}$ we see that the $\vec{n}(t)$ part of $(\vec{n}(t), a(t))$ is autonomous and is just the same as the three-dimensional case of the example following (30). The $a(t)$ part does not change except at the times when the $\vec{n}(t)$ part jumps, and if $\vec{n}(t)$ jumps from $\vec{n}$ to $\vec{n} \pm u_i$ then $a(t)$ changes sign with probability $n_i^2/|\vec{n}|^2$. It is clear from (33) that the $a(t)$ part of the process causes some cancellation, and in fact there is enough cancellation to give ergodicity for T_t. To see this we write

$$\widehat{T_t^*\mu}(\vec{n}) = \sum_{\vec{n}} [\hat{P}_{(\vec{n},1)}(\vec{n}(t) = \vec{m}, a(t) = 1) - \hat{P}_{(\vec{n},1)}(\vec{n}(t) = m, a(t) = -1)]\hat{\mu}(\vec{m}).$$

Thus if $\tau_0 \inf\{t: \vec{n}(t) = 0\}$ we have

$$\left|\widehat{T_t^*\mu}(\vec{n}) - \int_{\{\tau_0 < \infty\}} a(\tau_0)\, d\hat{P}_{(\vec{n},1)}\right| \leq \hat{P}_{(\vec{n},1)}(t < \tau_0 < \infty) + \sum_{\vec{m}\neq 0}\left|\hat{P}_{(\vec{n},1)}(\vec{n}(t) = m, a(t) = 1) - \hat{P}_{(\vec{n},1)}(\vec{n}(t) = m, a(t) = -1)\right|. \tag{34}$$

We prove ergodicity of T_t and identify the Fourier coefficients of its stationary measure by showing that the right side of (34) goes to zero as t goes to infinity.

The first term clearly goes to zero. To handle the second term let $\mathscr{F} = \mathscr{B}[\vec{n}(t): t \geqq 0]$, $N(t)$ be the number of jumps which the $\vec{n}(t)$ part of the process makes by time t, $n^{(j)}$ be the position of $\vec{n}(t)$ immediately after the jth jump, and d_j be i if the $j+1$st jump is in the direction $\pm u_i$. From our description of the process it is intuitively clear that

$$\hat{E}_{(\vec{n},a)}[a(t)\,|\,\mathscr{F}] = a \prod_{j=0}^{N(t)-1} \frac{|\vec{n}^{(j)}|^2 - 2(n_{d_j}^{(j)})^2}{|\vec{n}^{(j)}|^2}. \tag{35}$$

A rigorous proof of equation (35) requires some work. We will not prove it here. Now

$$\begin{aligned}
&|\hat{P}_{(\vec{n},1)}(\vec{n}(t) = m, a(t) = 1) - P_{(\vec{n},1)}(\vec{n}(t) = m, a(t) = -1)| \\
&= \left| \int_{\{\vec{n}(t)=m\}} E[a(t)|\mathscr{F}]\, d\hat{P}_{(\vec{n},1)} \right| \\
&\leqq \int_{\{\vec{n}(t)=m\}} \prod_{j=0}^{N(t)-1} \left| \frac{|\vec{n}^{(j)}|^2 - 2(n_{d_j}^{(j)})^2}{|\vec{n}^{(j)}|^2} \right| d\hat{P}_{(\vec{n},j)} \\
&\leqq \int_{\{\vec{n}(t)=m\}} 3^{-(M(t)-1)}\, d\hat{P}_{(\vec{n},1)},
\end{aligned}$$

where $M(t)$ is the number of visits of $\vec{n}(\cdot)$ to $\{(k, k, k): k \in Z\}$ by time t. Therefore the second term on the right side of (34) is bounded by

$$\int_{\{\vec{n}(t) \neq 0\}} 3^{-(M(t)-1)}\, dP_{(\vec{n},1)}.$$

Since

$$\hat{P}_{(\vec{n},1)}(\vec{n}(t) = 0 \text{ for all large } t \text{ or } \vec{n}(t) \text{ visits } \{(k,k,k): k \in Z\} \text{ infinitely often}) = 1,$$

we have the desired result.

References

1. K.-L. Chung and W. H. J. Fuchs, *On the distribution of sums of random variables*, Mem. Amer. Math. Soc. **6** (1951), 1–12. MR **12**, 722.

2. I. V. Girsanov, *An example of nonuniqueness of the solution of K. Itô's stochastic integral equation*, Teor. Verojatnost. i Primenen **7** (1962), 336–342.

3. R. A. Holley and D. W. Stroock, *Dual processes and their application to infinite interacting systems*, Advances in Math. (to appear).

4. K. Ichihara and H. Kunita, *A classification of second order degenerate elliptic operators and its probabilistic characterization*, Z. Wahrscheinlichkeitstheorie Verw. Gerbeite **30** (1974), 235–254.

5. N. V. Krylow, *On the selection of a Markov process from a Markov system of processes and the construction of quasi-diffusion processes*, Izv. Akad. Nauk SSSR Ser. Mat. **37** (1973), 691–708. MR **49** #4097.

6. D. W. Stroock and S. R. S. Varadhan, *Diffusion processes with continuous coefficients*. I, Comm. Pure Appl. Math. **22** (1969), 345–400. MR **40** #6641.

7. S. Watanabe and T. Yamada, *On the uniqueness of solutions of stochastic differential equations*. I, J. Math. Kyoto Univ. **11** (1971), no. 1, 155–167. MR **43** #4150.

8. W. K. Feller, *On boundaries and lateral conditions for the Kolmogorov equations*, Ann. of Math. (2) **65** (1957), 527–570. MR **19**, 892.

9. K. Itô, *Poisson point processes attached to Markov processes*, Proc. Sixth Berkeley Sympos. on Math. Statist. and Probability, vol. III, Univ. of California Press, Berkeley, 1972, pp. 225–240.

10. G. E. H. Reuter, *Denumerable Markov processes*. II, J. London Math. Soc. **34** (1959), 81–91. MR **21** #919.

UNIVERSITY OF COLORADO

UNIVERSITY COLLEGE OF SWANSEA

Proceedings of Symposia in Pure Mathematics
Volume 31, 1977

A GENERAL THEOREM OF REPRESENTATION FOR MARTINGALES

JEAN JACOD

Introduction. Since the celebrated theorem asserting that any martingale of a Wiener process is a stochastic integral with respect to this Wiener process, many similar "representation theorems" have been found: for some sort of diffusion processes, for Poisson processes, for general point processes, for processes with independent increments, and so on. Moreover these representation theorems turn out to be most useful in several application fields, including filtering and control theory.

The aim of this paper is to exhibit a common property of all these processes, which implies for them the "representation theorem for martingales".

More precisely we set a general martingale problem: A solution of such a problem is a probability measure. Then we show that a solution is extremal in the set of all solutions if and only if it satisfies a representation theorem, that is any martingale is a sum of stochastic integrals with respect to a family of a priori known martingales.

In particular when the solution is unique, it satisfies a representation theorem: This idea of proof was already used by Dellacherie [**4**] for Poisson and Wiener processes, and in [**12**] for general point processes.

As a corollary we obtain many of the known results of representation for martingales. We obtain also some new (as far as we know) results: For example any martingale of diffusion processes in a domain of $\boldsymbol{R}^d$ with a reflecting boundary is, under suitable conditions, stochastic integral with respect to d fundamental martingales; similarly, martingales of "diffusion processes with jumps" (in the sense of Stroock [**23**]) have a nice representation as stochastic integrals (however, the fundamental result of Motoo and Watanabe about martingales of a general Markov process is not a simple corollary of our theorem).

AMS (*MOS*) *subject classifications* (1970). Primary 60G45, 60H05; Secondary 60J60, 60J75.

The organization of this paper is as follows: In §1 we give the necessary preliminaries; this is somewhat lengthy and a reader interested only in continuous processes may skip §1.3. In §2 we state our martingale problem. §3 is devoted to the statement of the representation property (3.1), of the main theorem (3.2) and to its proof. At last, all examples have been gathered in §4.

1. Preliminaries.

1. *Some notations.* Let $(\Omega, \mathscr{F})$ be a measurable space on which is defined an increasing and right-continuous family $(\mathscr{F}_t)_{t\geqq 0}$ of sub-σ-algebras of $\mathscr{F}$ such that $\mathscr{F} = \bigvee \mathscr{F}_t$.

We shall consider only real-valued processes. For any process $X = (X_t)$ which is right-continuous and admits left-hand limits, we put $\Delta X_t = X_t - X_{t-}$.

We make frequent use of the books by Meyer [**17**] and Dellacherie [**3**], concerning well measurable and predictable processes. Let us just give here a definition which relies only upon the measurable structure $(\mathscr{F}_t)$.

DEFINITION. A *process* (X_t) is called *well-measurable* (resp. *predictable*) if the application $(\omega, t) \rightsquigarrow X_t(\omega)$ is measurable with respect to the σ-algebra $\mathscr{W}$ (resp. $\mathscr{P}$) of $\Omega \times [0, \infty[$ generated by the applications $(\omega, t) \rightsquigarrow Y(\omega, t)$ which are $\mathscr{F}_t$-measurable in ω and right-continuous (resp. left-continuous) in t. Recall that $\mathscr{P} \subset \mathscr{W}$.

On the other hand, let E be a universally measurable subset of a compact metric space, and $\mathscr{E}$ be the Borel σ-algebra of E. We put $\tilde{E} =]0, \infty[\times E$ and $\tilde{\mathscr{E}} = \mathscr{B}(]0, \infty[) \otimes \mathscr{E}$. We also define $\tilde{\Omega} = \Omega \times [0, \infty[\times E$, with the σ-algebras $\tilde{\mathscr{W}} = \mathscr{W} \otimes \mathscr{E}$ and $\tilde{\mathscr{P}} = \mathscr{P} \otimes \mathscr{E}$.

DEFINITION. A *random measure* is a positive transition measure $\mu(\omega; dt, dx)$ from $(\Omega, \mathscr{F})$ into $(\tilde{E}, \tilde{\mathscr{E}})$. A *random measure* μ is called *integer-valued* if it takes its values in $\boldsymbol{N}^* \cup \{+\infty\}$, and if $\mu(\omega; \{t\} \times E) \leqq 1$ identically.

Let μ be a random measure, and X be a nonnegative $\mathscr{F} \otimes \mathscr{B}([0, \infty[) \otimes \mathscr{E}$-measurable function on $\tilde{\Omega}$. Then one can define a new random measure $X \cdot \mu$ and an increasing process $X * \mu$ by

$$
(1) \qquad \begin{aligned} &(X \cdot \mu)(\omega; dt, dx) = \mu(\omega; dt, dx) X(\omega, t, x), \\ &(X * \mu)_t(\omega) = (X \cdot \mu)(\omega;]0, t] \times E) = \int_{]0,t] \times E} X(\omega, s, y) \mu(\omega; ds, dy). \end{aligned}
$$

DEFINITION. A *random measure* μ is called *well-measurable* (resp. *predictable*) if the process $X * \mu$ is well measurable (resp. predictable) for any $\tilde{\mathscr{W}}$-measurable (resp. $\tilde{\mathscr{P}}$-measurable) nonnegative function X on $\tilde{\Omega}$.

2. *Martingales and stochastic integrals.* Let P be a probability measure on $(\Omega, \mathscr{F})$. When two processes X and Y satisfy $X_t(\omega) = Y_t(\omega)$ for each $t \geqq 0$ and each ω not belonging to a P-null set, we write $X = Y(P)$ and we say that X and Y are P-equal.

The set of all processes A which are P-a.s. increasing and right-continuous and which satisfy $A_0 = 0$ P-a.s. and $A_t < \infty$ P-a.s. for each $t < \infty$ is denoted by $\mathscr{A}^+(P)$, and $\mathscr{A}(P) = \mathscr{A}^+(P) - \mathscr{A}^+(P)$. If $A \in \mathscr{A}(P)$ and if X is a measurable process, we define the process $X \cdot A$ by

$$
\begin{aligned} X \cdot A_t(\omega) &= \int_0^t X_s(\omega)\, dA_s(\omega) \quad \text{if} \int_0^t |X_s(\omega)|\, |dA_s(\omega)| < \infty, \\ &= +\infty \qquad\qquad\qquad\quad \text{if not.} \end{aligned}
$$

We write $\mathscr{V}_{\text{loc}}(P)$ for the set of all $A \in \mathscr{A}(P)$ which are *locally integrable*, i.e., for which there exists a sequence (T_n) of stopping times increasing P-a.s. to $+\infty$, such that $E(\int_{]0,T_n]} |dA_t|)$ is finite for each n. It is known that any *predictable* element of $\mathscr{A}(P)$ belongs to $\mathscr{V}_{\text{loc}}(P)$ (**[12]** or **[19]**).

For all the results concerning martingales and stochastic integrals, we refer to Meyer **[17]**, Doléans-Dade and Meyer **[6]**, or Meyer **[19]**.

$\mathscr{M}(P)$ is the set of all right-continuous *uniformly integrable* $(\Omega, \mathscr{F}_t, P)$-martingales, with the convention that we identify two martingales which are P-equal. Let $\mathscr{M}^2(P)$ be the set of *square-integrable martingales which are null at the origin*, i.e., elements $M \in \mathscr{M}(P)$ such that $M_0 = 0$ and $\sup_{(t)} E(M_t^2) < \infty$. Let $\mathscr{M}_{\text{loc}}(P)$ (resp. $\mathscr{M}^2_{\text{loc}}(P)$) be the set of all *local martingales* (resp. *locally square-integrable martingales*), that is processes M for which there exists a sequence (T_n) of stopping times increasing P-a.s. to $+\infty$, such that each process $(M_t^{T_n}) = (M_{T_n \wedge t})$ belongs to $\mathscr{M}(P)$ (resp. $\mathscr{M}^2(P)$). If $M \in \mathscr{M}_{\text{loc}}(P)$ (resp. $\mathscr{M}^2_{\text{loc}}(P)$) and if T is a stopping time such that $M^T \in \mathscr{M}(P)$ (resp. $\mathscr{M}^2(P)$), we say that T *reduces* M.

The following facts are well known: if $M, N \in \mathscr{M}^2_{\text{loc}}(P)$ there exists a unique (up to a modification on a P-null set) predictable process $\langle M, N \rangle \in \mathscr{V}_{\text{loc}}(P)$ such that $MN - \langle M, N \rangle \in \mathscr{M}_{\text{loc}}(P)$, and moreover $\langle M, M \rangle \in \mathscr{A}^+(P)$. If $M \in \mathscr{M}_{\text{loc}}(P)$ satisfies $M_0 = 0$ and $|\Delta M| \leq b$ except on a P-null set, for a finite b, then $M \in \mathscr{M}^2_{\text{loc}}(P)$. Each $M \in \mathscr{M}_{\text{loc}}(P)$ admits a unique decomposition $M = M^c + M^d$ where M^c is a continuous element of $\mathscr{M}^2_{\text{loc}}(P)$, and M^d is a *compensated sum of jumps*, i.e., an element of $\mathscr{M}_{\text{loc}}(P)$ satisfying $M^d N \in \mathscr{M}_{\text{loc}}(P)$ for each continuous $N \in \mathscr{M}^2_{\text{loc}}(P)$; M^c is called the *continuous part* of M.

If $M \in \mathscr{M}^2_{\text{loc}}(P)$ we denote by $L^2_{\text{loc}}(M, P)$ the set of all predictable processes X such that $X^2 \cdot \langle M, M \rangle \in \mathscr{V}_{\text{loc}}(P)$; then if $X \in L^2_{\text{loc}}(M, P)$ one can define the *stochastic integral* $X \cdot M$ as being the only one element of $\mathscr{M}^2_{\text{loc}}(P)$ such that $\langle X \cdot M, N \rangle = X \cdot \langle M, N \rangle$ for every $N \in \mathscr{M}^2_{\text{loc}}(P)$.

3. *Random measures and stochastic integrals.* To each random measure we associate a positive measure M_μ^P on $(\tilde{\Omega}, \tilde{\mathscr{W}})$ by the formula $M_\mu^P(X) = E(X * \mu_\infty)$, where X is any nonnegative $\tilde{\mathscr{W}}$-measurable function on $\tilde{\Omega}$. We say that M_μ^P is $\tilde{\mathscr{P}}$-*σ-finite* if its restriction to $(\tilde{\Omega}, \tilde{\mathscr{P}})$ is σ-finite, and we denote by $\mathscr{K}(\mu, P)$ the set of all $\tilde{\mathscr{W}}$-measurable functions X such that $M^P_{|X| \cdot \mu}$ is $\tilde{\mathscr{P}}$-σ-finite. When M_μ^P is $\tilde{\mathscr{P}}$-σ-finite and when $X \in \mathscr{K}(\mu, P)$, we may define the *conditional expectation* $M_\mu^P(X | \tilde{\mathscr{P}})$ as being any $\tilde{\mathscr{P}}$-measurable version of the Radon-Nikodým derivative of the restriction of the measure $X \cdot M_\mu^P$ to $(\tilde{\Omega}, \tilde{\mathscr{P}})$, with respect to the restriction of M_μ^P to $(\tilde{\Omega}, \tilde{\mathscr{P}})$.

Here is a fundamental result **[12]**: *If μ is a random measure such that M_μ^P is $\tilde{\mathscr{P}}$-σ-finite, there exists a unique* (up to a modification on a P-null set) *predictable random measure ν such that the restrictions of M_μ^P and M_ν^P to $(\tilde{\Omega}, \tilde{\mathscr{P}})$ are equal.* The random measure ν is called the *dual predictable projection of* μ. Moreover when μ is well-measurable, then for each nonnegative $\tilde{\mathscr{P}}$-measurable function X such that $X * \mu \in \mathscr{V}_{\text{loc}}(P)$, then $X * \mu - X * \nu \in \mathscr{M}_{\text{loc}}(P)$.

EXAMPLE. Let Y be an adapted right-continuous and left-hand limited process, and let $E = \boldsymbol{R}$. Then the formula

$$\mu(\omega; dt, dx) = \sum_{(s)} 1_{\{\Delta Y_s(\omega) \neq 0\}} \, \varepsilon_{(s, \Delta Y_s(\omega))}(dt, dx) \tag{2}$$

is a well-measurable integer-valued random measure μ, and M_μ^P is $\tilde{\mathscr{P}}$-σ-finite for any probability measure P. The dual predictable projection of μ is known under the name of "Lévy system" of Y.

From now on, we suppose that μ is a *well-measurable integer-valued random measure such that M_μ^P is $\tilde{\mathscr{P}}$-σ-finite*. Let ν be its dual predictable projection. If $a_t(\omega) = \nu(\omega; \{t\} \times E)$ one knows **[13]** that one may, and will, choose a version of ν for which $a_t(\omega) \leqq 1$ identically.

Now we proceed to the construction of stochastic integrals with respect to $(\mu - \nu)$. As this is less classical than stochastic integrals with respect to martingales, we will go into more details. The reference is Jacod **[13]** (and also Skorokhod **[21]** when $a_t = 0$).

For notational convenience we put $J = \{(\omega, t): a_t(\omega) > 0\}$, which belongs to $\mathscr{P}$. For each function U on $\tilde{\Omega}$ we define two increasing processes:

$$\beta_t(U) = |U 1_{J^c}| * \nu_t + \sum_{s \leqq t} 1_J(s) \left| \int_E \mu(\{s\}, dx) U(s, x) - \int_E \nu(\{s\}, dx) U(s, x) \right|,$$

$$B_t(U) = (U^2 1_{J^c}) * \nu_t + \sum_{s \leqq t} 1_J(s) \left[\int_E \nu(\{s\}, dx) U^2(s, x) - \left[\int_E \nu(\{s\}, dx) U(s, x) \right]^2 \right],$$

with the convention that $\beta_t(U) = +\infty$ (resp. $B_t(U) = +\infty$) if one of the above integrals is not well defined, or if we arrive in some place to an expression "$\infty - \infty$". Let

$$\begin{aligned} \mathscr{G}^1_{\mathrm{loc}}(\mu, P) &= \{U: \tilde{\mathscr{P}}\text{-measurable}, \beta(U) \in \mathscr{V}_{\mathrm{loc}}(P)\}, \\ \mathscr{G}^2_{\mathrm{loc}}(\mu, P) &= \{U: \tilde{\mathscr{P}}\text{-measurable}, B(U) \in \mathscr{V}_{\mathrm{loc}}(P)\}, \\ \mathscr{G}_{\mathrm{loc}}(\mu, P) &= \{U = U_1 + U_2, U_1 \in \mathscr{G}^1_{\mathrm{loc}}(\mu, P), U_2 \in \mathscr{G}^2_{\mathrm{loc}}(\mu, P)\}. \end{aligned}$$

Then for each $U \in \mathscr{G}_{\mathrm{loc}}(\mu, P)$ one can define an element $U * (\mu - \nu)$ of $\mathscr{M}_{\mathrm{loc}}(P)$, with the following properties:

(i) $U \in \mathscr{G}^1_{\mathrm{loc}}(\mu, P)$ implies $U * (\mu - \nu) \in \mathscr{M}_{\mathrm{loc}}(P) \cap \mathscr{V}_{\mathrm{loc}}(P)$.

(ii) $U \in \mathscr{G}^2_{\mathrm{loc}}(\mu, P)$ implies $U * (\mu - \nu) \in \mathscr{M}^2_{\mathrm{loc}}(P)$ and

$$\langle U * (\mu - \nu), U * (\mu - \nu) \rangle = B(U).$$

(iii) $U * (\mu - \nu)$ is a compensated sum of jumps, and

$$\Delta[U * (\mu - \nu)]_t = \int_E \mu(\{t\}, dx) U(t, x) - \int_E \nu(\{t\}, dx) U(t, x).$$

(iv) When $|U| * \mu \in \mathscr{V}_{\mathrm{loc}}(P)$, we have $U * (\mu - \nu) = U * \mu - U * \nu$ (where $U * \mu$ and $U * \nu$ are two elements of $\mathscr{A}(P)$ defined by an immediate extension of (1)).

We end this section by a result of **[13]** which will be used later.

THEOREM 1. (a) *If $M \in \mathscr{M}_{\mathrm{loc}}(P)$, then $\Delta M \in \mathscr{K}(\mu, P)$ (by ΔM we mean the function $\Delta M(\omega, t, x) = \Delta M_t(\omega)$).*

(b) *Let $M \in \mathscr{M}_{\mathrm{loc}}(P)$, $U = M_\mu^P(\Delta M | \tilde{\mathscr{P}})$ and*

$$W(t, x) = U(t, x) + 1_{\{a_t < 1\}} \frac{1}{1 - a_t} \int_E \nu(\{t\}, dy) U(t, y). \tag{3}$$

Then $W \in \mathscr{G}_{\mathrm{loc}}(\mu, P)$ and $N' = N - W * (\mu - \nu)$ satisfies $M_\mu^P(\Delta N' | \tilde{\mathscr{P}}) = 0$.

(c) *If in addition $M \in \mathscr{M}^2_{\mathrm{loc}}(P)$, then $W \in \mathscr{G}^2_{\mathrm{loc}}(\mu, P)$ and we have $\langle N', W * (\mu - \nu) \rangle = 0$.*

2. A martingale problem. In all the sequel we fix $(\Omega, \mathscr{F})$ and the family $(\mathscr{F}_t)$, as well as the space $(E, \mathscr{E})$. We also fix a well-measurable integer-valued random measure μ and a family $X = (X^i, i \in I)$ indexed by an arbitrary set I, consisting in $(\mathscr{F}_t)$-adapted, right-continuous and left-hand limited processes. At last we fix a sub-σ-algebra $\mathscr{F}_0^0$ of $\mathscr{F}_0$, which will play the role of the "initial condition" (usually, $\mathscr{F}_0^0$ is the σ-algebra generated by the random variables X_0^i).

In order to formulate our martingale problem, we consider a set $\mathscr{Q}$ whose points are quintuplets $\boldsymbol{Q} = (Q, \nu, W, A, B)$ consisting in:

a probability measure Q on $(\Omega, \mathscr{F}_0^0)$,

a predictable random measure ν,

a family $W = (W)^i_{i \in I}$ of $\mathscr{P}$-measurable functions on $\tilde{\Omega}$,

a family $A = (A^i)_{i \in I}$ of predictable, right-continuous and left-hand limited processes,

a family $B = (B^{ij})_{i,j \in I}$ of continuous adapted processes.

DEFINITION. A *probability measure* P on $(\Omega, \mathscr{F})$ is *a solution of the martingale problem* $(\mu, X, \mathscr{F}_0^0, \mathscr{Q})$ if there exists $\boldsymbol{Q} = (Q, \nu, W, A, B) \in \mathscr{Q}$ such that:

(i) Q is the restriction of P to $(\Omega, \mathscr{F}_0^0)$.

(ii) M_μ^P is $\tilde{\mathscr{P}}$-σ-finite and ν is a version of the dual predictable projection of μ for P.

(iii) For each $i \in I$, $A^i \in \mathscr{V}_{\mathrm{loc}}(P)$, $W^i \in \mathscr{G}_{\mathrm{loc}}(\mu, P)$, and the process

$$M_t^i = X_t^i - X_0^i - W^i * (\mu - \nu)_t - A_t^i \tag{4}$$

is P-a.s. continuous and belongs to $\mathscr{M}^2_{\mathrm{loc}}(P)$.

(iv) For each $i, j \in I$, we have $B^{ij} = \langle M^i, M^j \rangle$.

In many (but not all) cases, the set $\mathscr{Q}$ consists in only one point. Most of the "usual" martingale problems, including construction of diffusion processes, diffusion processes with reflection, processes with independent increments, are of the above type: The reader interested in applications can look immediately at §4, where many examples are displayed.

As μ, X and $\mathscr{F}_0^0$ are fixed throughout all the paper, we shall denote by $\mathscr{S}(\mathscr{Q})$ the set of all solutions of the martingale problem $(\mu, X, \mathscr{F}_0^0, \mathscr{Q})$. If $P \in \mathscr{S}(\mathscr{Q})$ and if $\boldsymbol{Q} \in \mathscr{Q}$ satisfies conditions (i) — (iv) above, we say that $\boldsymbol{Q}$ is *associated* to P. The next proposition shows that the $\boldsymbol{Q}$ associated to P, as well as the local martingales M^i defined by (4), are essentially unique.

PROPOSITION 1. *Let* $P \in \mathscr{S}(\mathscr{Q})$. *Suppose that* $\boldsymbol{Q} = (Q, \nu, W, A, B)$ *and* $\boldsymbol{Q}' = (Q', \nu', W', A', B')$ *are associated to P, and let* M^i *(resp.* M'^i*) be associated to* $\boldsymbol{Q}$ *(resp.* $\boldsymbol{Q}'$*) by* (4). *Then* $Q = Q'$, $\nu = \nu'$ *except on a P-null set,* $A^i = A'^i(P)$, $M^i = M'^i(P)$ *and* $B^{ij} = B'^{ij}(P)$ *for each* $i, j \in I$.

PROOF. $Q = Q'$ is evident, and $\nu = \nu'$ follows from the uniqueness of the dual predictable projection of μ. We have

$$M'^i - M^i = (W^i - W'^i) * (\mu - \nu) + (A^i - A'^i),$$

which implies $A^i - A'^i \in \mathscr{M}_{\mathrm{loc}}(P)$. But from Dellacherie [3], a local martingale which is predictable and belongs to $\mathscr{V}_{\mathrm{loc}}(P)$ is 0. Thus $A^i = A'^i(P)$. Moreover $M'^i - M^i$ is continuous and $(W^i - W'^i) * (\mu - \nu)$ is a compensated sum of jumps, so $M'^i = M^i(P)$ and it follows that $B^{ij} = B'^{ij}(P)$. $\square$

REMARKS. (1) From condition (iv) we see that the B^{ij}'s are not only continuous; actually they also satisfy

$$(5)\qquad \begin{array}{ll} \forall\, i, j \in I, & B^{ij} = B^{ji}(P), \\ \forall\, J \subset I,\ J \text{ finite},\ \forall\, a_i \in \boldsymbol{R}, & \sum_{i,j \in J} a_i a_j B^{ij} \in \mathscr{A}^+(P). \end{array}$$

(2) In some respects it would be more natural to state a more general martingale problem: namely to start with a family $(\mu^i)_{i \in I}$ of random measures, ν being also a family $(\nu^i)_{i \in I}$ of predictable random measures, (4) being replaced by $M^i = X^i - X^i_0 - W^i * (\mu^i - \nu^i) - A^i$, where $W^i \in \mathscr{G}_{\text{loc}}(\mu^i, P)$. But the main result of this paper would not hold true in this more general context (all the applications we have in mind fall into the scope of a unique μ; in addition when I is countable one can always aggregate the measures μ^i, to give a new integer-valued random measure μ on the space E^I, which is again a universally measurable subset of a compact metric space).

(3) In most applications, martingale problems are stated in a different form: we consider a set $\mathscr{Q}'$ of quadruplets (Q, ν, A, B) (dropping the "W" in the definition above); then we say that P is a solution if (i) and (ii) hold, if (iii) reads as follows: $A^i \in \mathscr{V}_{\text{loc}}(P)$ and there exists $W^i \in \mathscr{G}_{\text{loc}}(\mu, P)$ such that M^i defined by (4) is a continuous element of $\mathscr{M}_{\text{loc}}(P)$, and (iv) holds. Such a martingale problem is also clearly a martingale problem in our sense (with $\mathscr{Q}$ being the set of all (Q, ν, W, A, B) such that $(Q, \nu, A, B) \in \mathscr{Q}'$ and that $W = (W^i)$ is any family of $\tilde{\mathscr{P}}$-measurable functions). However the wider generality of our approach is somewhat illusory, as it will be seen in Example 4.2(b) (Proposition 4),

3. The main theorem.

1. *The property of representation of martingales.* Let $P \in \mathscr{S}(\mathscr{Q})$, $\boldsymbol{Q} = (Q, \nu, W, A, B)$ be associated to P, and M^i be defined by (4).

DEFINITION. We say that P enjoys *the property of representation for martingales* (and we write $P \in \mathscr{RM}$) if any $N \in \mathscr{M}^2_{\text{loc}}(P)$ satisfying $\langle N, M^i \rangle = 0$ for each $i \in I$ and $\langle N, U * (\mu - \nu) \rangle = 0$ for each $U \in \mathscr{G}^2_{\text{loc}}(\mu, P)$ is equal to 0.

Equivalently, we can say that $P \in \mathscr{RM}$ *if and only if the stable subspace of* $\mathscr{M}^2_{\text{loc}}(P)$ *generated by the* $(M^i, i \in I)$ *and the* $(U * (\mu - \nu), U \in \mathscr{G}^2_{\text{loc}}(\mu, P))$ *is exactly* $\mathscr{M}^2_{\text{loc}}(P)$: Following Meyer **[18]**, **[19]** and with the necessary changes due to the fact that we use $\mathscr{M}^2_{\text{loc}}(P)$ instead of $\mathscr{M}^2(P)$, we say that a subspace $\mathscr{L}$ of $\mathscr{M}^2_{\text{loc}}(P)$ is *stable* if $H \cdot M \in \mathscr{L}$ whenever $M \in \mathscr{L}$ and $H \in L^2_{\text{loc}}(M, P)$, and if $\mathscr{L} \cap \mathscr{M}^2(P)$ is closed in $\mathscr{M}^2(P)$ under the norm $\|M\| = (\sup_{(t)} E(M_t^2))^{1/2}$.

We can give still another, perhaps more "concrete", formulation of the property of representation for martingales:

PROPOSITION 2. *Condition $P \in \mathscr{RM}$ holds if and only if for any $N \in \mathscr{M}_{\text{loc}}(P)$ (resp. $\mathscr{M}^2_{\text{loc}}(P)$) there exist $U \in \mathscr{G}_{\text{loc}}(\mu, P)$ (resp. $\mathscr{G}^2_{\text{loc}}(\mu, P)$), an increasing sequence (J_n) of finite subsets of I, and elements $u^i_n \in L^2_{\text{loc}}(M^i, P)$, such that for each stopping time T reducing M,*

$$(6)\qquad N_T = N_0 + U * (\mu - \nu)_T + \lim_{(n)} \sum_{i \in J_n} (u^i_n \cdot M^i)_T$$

(*where the limit is to be taken in* $L^2(\Omega, \mathscr{F}, P)$).

REMARK. Let us assume I is finite. Under the condition $B^{ij} = 0\,(P)$ whenever $i \neq j$, property $P \in \mathscr{R}\mathscr{M}$ has a nicer form: Namely $P \in \mathscr{R}\mathscr{M}$ if and only if any $N \in \mathscr{M}_{\mathrm{loc}}(P)$ can be written as

$$N = N_0 + U * (\mu - \nu) + \sum_{i \in I} u^i \cdot M^i \tag{7}$$

where $u^i \in L^2_{\mathrm{loc}}(M^i, P)$ and $U \in \mathscr{G}_{\mathrm{loc}}(\mu, P)$. Without the above assumption, (7) usually no longer holds, because the set $\{\sum_{i \in I} u^i \cdot M^i \colon u^i \in L^2(M^i, P)\}$ is not closed in $\mathscr{M}^2(P)$. However (7) still holds under the following condition: If $B = \sum_{i \in I} B^{ii}$, there exist versions λ^{ij} of dB^{ij}/dB such that

$$\inf_{(\omega \in \Omega,\, (x_i) \in \boldsymbol{R}^d \setminus \{0\})} \frac{1}{\|x\|^2} \sum_{i,j \in I} x_i x_j \lambda^{ij}(\omega) > 0.$$

Before proceeding to the proof of Proposition 2, we give a lemma which we will use again later. We shall say that an element N of $\mathscr{M}_{\mathrm{loc}}(P)$ is *trivial* if $N_t = N_0$ P-a.s. for each t.

LEMMA 1. *If there exists a nontrivial* $N \in \mathscr{M}_{\mathrm{loc}}(P)$ *such that*

$$M^P_\mu(\Delta N \mid \tilde{\mathscr{P}}) = 0, \qquad \langle N^c, M^i \rangle = 0 \ \forall\, i \in I, \tag{8}$$

then there exists a nontrivial $N' \in \mathscr{M}^2_{\mathrm{loc}}(P)$ *satisfying* (8) *and* $|\Delta N'| \leqq 1$.

PROOF. First of all, if $N^c \neq 0$ we put $N' = N^c$. Suppose now $N^c = 0$ and $M^P_\mu(|\Delta N|) > 0$. As $M^P_\mu(\Delta N \mid \tilde{\mathscr{P}}) = 0$, the completions of $\tilde{\mathscr{P}}$ and of $\tilde{\mathscr{W}}$ with respect to the measure M^P_μ are different. Thus there exists $A \in \tilde{\mathscr{W}}$ such that $M^P_\mu(A) < \infty$ and $M^P_\mu(1_A \mid \tilde{\mathscr{P}})$ is not M^P_μ-a.s. equal to 1_A. Let $V = 1_A - M^P_\mu(1_A \mid \tilde{\mathscr{P}})$: As $|V| * \mu_\infty < \infty$ P-a.s., one can define $N' = V * \mu$ without ambiguity by an immediate extension of (1). N' is adapted and for each stopping time T we have $E(N'_T) = M^P_\mu(V 1_{\{t \leq T\}}) = 0$ because $M^P_\mu(V \mid \tilde{\mathscr{P}}) = 0$. Therefore $N' \in \mathscr{M}(P)$ and it satisfies all the requirements.

Now suppose $N^c = 0$ and $M^P_\mu(|\Delta N|) = 0$. Let us assume the existence of a totally inaccessible stopping time T such that $\Delta N_T \neq 0$ on $\{T < \infty\}$, and put $C_t = 1_{\{T \leq t\}}$: If C' is the dual predictable projection [3] of C, then $N' = C - C'$ is a solution to our problem. At last when no such T as above does exist, then there exists a predictable stopping time S with $0 < E(|\Delta N_S|) < \infty$. Let $D = \{S < \infty,\, \mu(\{S\} \times E) = 1\}$ and $\mathscr{H} = \{A \in \mathscr{F}_S,\, P(A \mid \mathscr{F}_{S-}) = 1_A \text{ on } D\}$: $\mathscr{H}$ is a σ-algebra containing $\mathscr{F}_{S-}$. From $E(\Delta N_S \mid \mathscr{F}_{S-}) = 0$ and $\Delta N_S = 0$ on D, we see that ΔN_S is $\mathscr{H}$-measurable, and the completions of $\mathscr{H}$ and of $\mathscr{F}_{S-}$ with respect to P are different. Thus we can choose $A \in \mathscr{H}$ such that $P(A \mid \mathscr{F}_{S-})$ is not P-a.s. equal to 1_A, and $N'_t = 1_{\{S \leq t\}}[1_A - P(A \mid \mathscr{F}_{S-})]$ is a solution. □

PROOF OF PROPOSITION 2. Let $\mathscr{L}$ be the stable subspace generated by $(M^i, i \in I)$. Then $P \in \mathscr{R}\mathscr{M}$ if and only if (i) any compensated sum of jumps $N^d \in \mathscr{M}^2_{\mathrm{loc}}(P)$ such that $\langle N^d, U * (\mu - \nu) \rangle = 0$ for each $U \in \mathscr{G}^2_{\mathrm{loc}}(\mu, P)$ is 0, and (ii) any continuous $N^c \in \mathscr{M}^2_{\mathrm{loc}}(P)$ is in $\mathscr{L}$.

But Theorem 1 immediately implies that (i) holds if and only if each $N^d \in \mathscr{M}^2_{\mathrm{loc}}(P)$ is $N^d = N^d_0 + U * (\mu - \nu)$ for a suitable $U \in \mathscr{G}^2_{\mathrm{loc}}(\mu, P)$. On the other hand $\mathscr{L} \cap \mathscr{M}^2(P)$ is the closure in $\mathscr{M}^2(P)$ of the set $\{\sum_{i \in J} u^i \cdot M^i \colon u^i \in L^2(M^i, P),\, J \text{ finite}\}$. Thus for any $N \in \mathscr{L}$ there exist an increasing sequence (J_n) of finite subsets of I and elements u^i_n of $L^2_{\mathrm{loc}}(M^i, P)$ such that, for each stopping time T reducing N, N_T is the limit in $L^2(\Omega, \mathscr{F}, P)$ of $\sum_{i\in J_n}(u_n^i \cdot M^i)_T$. Consequently (ii) holds if and only if each continuous $N^c \in \mathscr{M}^2_{\mathrm{loc}}(P)$ can be written like the second half of the right-hand side of (6).

To end the proof we have to show that when $P \in \mathscr{RM}$, any compensated sum of jumps $N^d \in \mathscr{M}_{\mathrm{loc}}(P)$ is $N^d = N^d_0 + U * (\mu - \nu)$ for some $U \in \mathscr{G}_{\mathrm{loc}}(\mu, P)$. If not, Theorem 1 implies the existence of a nontrivial $N \in \mathscr{M}_{\mathrm{loc}}(P)$ satisfying (8). Therefore if N' is the local martingale constructed in Lemma 1, another use of Theorem 1 shows that $\langle N', U * (\mu - \nu)\rangle = 0$ for any $U \in \mathscr{G}^2_{\mathrm{loc}}(\mu, P)$. As $\langle N'^c, M^i\rangle = 0$, this contradicts the assumption $P \in \mathscr{RM}$. □

3. *Statement of the theorem.* For any probability measure P on $(\Omega, \mathscr{F})$ we denote by $\mathscr{N}^P$ the set of all P-null sets belonging to the completion of $\mathscr{F}$ with respect to P.

THEOREM 2. *Let P be a solution of the martingale problem $(\mu, X, \mathscr{F}^0_0, \mathscr{Q})$.*

(i) *If P is extremal in the set $\mathscr{S}(\mathscr{Q})$ we have both*

(a) $\mathscr{F}^0_0 \vee \mathscr{N}^P = \mathscr{F}_0 \vee \mathscr{N}^P$,

(b) $P \in \mathscr{RM}$.

(ii) *When $\mathscr{Q}$ consists in only one point, then the set of solutions $\mathscr{S}(\mathscr{Q})$ is convex. If in addition both* (a) *and* (b) *hold, then P is extremal in $\mathscr{S}(\mathscr{Q})$.*[1]

For the above general formulation of Theorem 2, thanks are due to Marc Yor: Without him it would have been assumed in part (i) uniqueness of the solution P in $\mathscr{S}(\mathscr{Q})$ instead of extremality.

4. *Proof of Theorem* 2. In this paragraph we have $P \in \mathscr{S}(\mathscr{Q})$, $\boldsymbol{Q} = (Q, \nu, W, A, B)$ is associated to P and M^i is defined by (4).

We begin by an auxiliary result. Let P' be another probability measure on $(\Omega, \mathscr{F})$, absolutely continuous with respect to P (one writes $P' \ll P$). Put $Z_\infty = dP'/dP$ and $Z_t = E(Z_\infty | \mathscr{F}_t)$, thus defining $Z = (Z_t) \in \mathscr{M}(P)$. We consider the restriction Q' of P' to $(\Omega, \mathscr{F}^0_0)$, and the set $\mathscr{Q}' = \{\boldsymbol{Q}'\}$, where $\boldsymbol{Q}' = (Q', \nu, W, A, B)$.

Here comes a classical result (for a proof, see for example Van Schuppen and Wong **[24]**).

LEMMA 2. *Let $R_n = \inf(t\colon Z_t \leqq 1/n)$. Then $M \in \mathscr{M}_{\mathrm{loc}}(P')$ if and only if $(MZ)^{R_n} \in \mathscr{M}_{\mathrm{loc}}(P)$ for each n.*

The next lemma is a very weakened form of Girsanov's theorem for a general martingale problem. One can find a stronger form (in a slightly less general context) in **[14]**.

LEMMA 3. *$P' \in \mathscr{S}(\mathscr{Q}')$ if and only if Z satisfies* (8).

PROOF. We break the proof into several steps (for similar proofs, see **[14]**, and also **[24]**).

(a) At first we recall from Dellacherie **[3]** that any well-measurable (resp. predictable) increasing process C satisfies

$$E'(C_\infty) = E(Z_\infty C_\infty) = E(Z \cdot C_\infty) \qquad (\text{resp. } = E(Z_- \cdot C_\infty)).$$

[1] A result of the same kind (also related to **[14]**) has been proven independently by R. Lipcer, in case $\mu = 0$, I consists in one element and X is continuous (*Representation of martingales*, 1976 (preprint)).

For any $D \in \tilde{\mathscr{P}}$, application of these facts to $C = 1_D * \mu$ and $C = 1_D * \nu$ yields $M_\mu^{P'}(D) = M_\mu^P(Z1_D)$ and $M_\nu^{P'}(D) = M_\nu^P(Z_- 1_D)$.

Therefore if $P' \in \mathscr{S}(\mathscr{Q}')$ we have $M_\mu^P(\Delta Z 1_D) = 0$ for each $D \in \tilde{\mathscr{P}}$, which implies $M_\mu^P(\Delta Z | \tilde{\mathscr{P}}) = 0$. Conversely if $M_\mu^P(\Delta Z | \tilde{\mathscr{P}}) = 0$ we see that $M_\mu^{P'}(D) = M_\mu^P(Z_- 1_D) = M_\nu^P(Z_- 1_D)$ for each $D \in \tilde{\mathscr{P}}$: It follows easily that $M_\mu^{P'}$ is $\tilde{\mathscr{P}}$-σ-finite and that ν is the dual predictable projection of μ for P'.

(b) From $P' \ll P$ we deduce easily $\mathscr{A}(P) \subset \mathscr{A}(P')$: Therefore any predictable element of $\mathscr{V}_{\text{loc}}(P)$ belongs to $\mathscr{V}_{\text{loc}}(P')$, and in particular $A^i \in \mathscr{V}_{\text{loc}}(P')$. It is shown in [**13**, Proposition (5-3)] that $W^i \in \mathscr{G}_{\text{loc}}(\mu, P)$ if and only if a certain predictable increasing process $C(\nu, W^i)$ depending only upon W^i and ν is in $\mathscr{V}_{\text{loc}}(P)$. Thus $C(\nu, W^i)$ is in $\mathscr{V}_{\text{loc}}(P')$ and, ν being the dual predictable projection of μ for P' also, we have $W^i \in \mathscr{G}_{\text{loc}}(\mu, P')$. Then one can define stochastic integrals $W^i * (\mu - \nu)$ either for P or for P', and these two integrals coincide because they are obtained by limiting procedures in L^2 and L^1 (cf. [**13**]).

(c) Let us recall the following consequence of generalized Itô's formula [**6**]. If $M, M' \in \mathscr{M}_{\text{loc}}(P)$ and if M is continuous and satisfies $M_0 = 0$, then

$$M_t M'_t = (M_- \cdot M')_t + (M'_- \cdot M)_t + \langle M, M'^c \rangle_t. \tag{9}$$

Now assume $P' \in \mathscr{S}(\mathscr{Q}')$: Then $M^i \in \mathscr{M}_{\text{loc}}(P')$ and from Lemma 2 we have $(M^i Z)^{R_n} \in \mathscr{M}_{\text{loc}}(P)$ for each n; on the other hand, (9) applied to $M = M^i$ and $M' = Z$ yields $M^i Z - \langle M^i, Z^c \rangle \in \mathscr{M}_{\text{loc}}(P)$. Putting together these two results, we obtain $\langle M^i, Z^c \rangle_t = 0$ for $t \leqq R_n$. But Z being a nonnegative martingale, we have $Z_t = 0$ if $t \geqq \lim_{(n)} R_n$: therefore $\langle M^i, Z^c \rangle = 0$, and we have finished the proof of the necessary condition.

Conversely let us assume that $\langle M^i, Z^c \rangle = 0$ for each $i \in I$. As above we obtain from (9) that $M^i Z \in \mathscr{M}_{\text{loc}}(P)$, and Lemma 2 implies $M^i \in \mathscr{M}_{\text{loc}}(P')$. Now if $M = M^i M^j - B^{ij}$ we see from (9) again that $M = M^i_- \cdot M^j + M^j_- \cdot M^i$. Applying (9) once more with $M' = Z$, we see that $MZ - \langle M, Z^c \rangle \in \mathscr{M}_{\text{loc}}(P)$. But

$$\langle M, Z^c \rangle = M^i_- \cdot \langle M^j, Z^c \rangle + M^j_- \cdot \langle M^i, Z^c \rangle = 0$$

and $MZ \in \mathscr{M}_{\text{loc}}(P)$. Therefore Lemma 2 implies $M \in \mathscr{M}_{\text{loc}}(P')$, and the sufficient condition is proven. □

PROOF OF THEOREM 2. Let us begin with part (i): if (a) does not hold, there exists $A \in \mathscr{F}_0$ such that $P(A | \mathscr{F}_0^0)$ is not P-a.s. equal to 1_A. Let us put

$$Z_t = 1 + \tfrac{1}{2}[1_A - P(A | \mathscr{F}_0^0)],$$

which defines a (trivial) element $Z \in \mathscr{M}(P)$. It is clear that $0 \leqq Z_\infty \leqq 4$, $E(Z_\infty) = 1$, Z_∞ is not P-a.s. equal to 1, and Z satisfies (8).

If condition (b) is unsatisfied there exists a nontrivial N in $\mathscr{M}^2_{\text{loc}}(P)$ with $\langle N^c, M^i \rangle = 0$ for each $i \in I$ and $\langle N, U * (\mu - \nu) \rangle = 0$ for each $U \in \mathscr{G}^2_{\text{loc}}(\mu, P)$. Theorem 1 implies that $M_\mu^P(\Delta N | \tilde{\mathscr{P}}) = 0$ and N satisfies (8). From Lemma 1 there exists $N' \in \mathscr{M}^2_{\text{loc}}(P)$ satisfying (8) and $|\Delta N'| \leqq 1$. Put

$$Z'_t = \exp(N'_t - \tfrac{1}{2} \langle N'^c, N'^c \rangle_t) \prod_{s \leqq t} [(1 + \Delta N'_s) e^{-\Delta N'_s}].$$

This is the exponential formula of Doléans-Dade [**5**] and Z' is the only local martingale, solution of the equation

(10) $$Z'_t = 1 + (Z'_- \cdot N')_t.$$

From $|\Delta N'| \leqq 1$ we get $Z'_t \geqq 0$. If $T = \inf(t\colon Z'_t \geqq 2)$, we have $Z'_{t\wedge T} \leqq 4$ because $Z'_t = Z'_{t-}(1 + \Delta N'_t)$. Therefore $Z = Z'^T$ is nontrivial and belongs to $\mathscr{M}(P)$. At last, using equation (10), it is an easy job to check that Z' and Z satisfy (8).

At this point we have shown that, when either (a) or (b) is not satisfied, there exists $Z \in \mathscr{M}(P)$ such that $E(Z_0 | \mathscr{F}^0_0) = 1$, $0 \leqq Z_\infty \leqq 4$, $E(Z_\infty) = 1$, Z_∞ is not P-a.s. equal to 1 and Z satisfies (8). If $Z^1 = (4 + Z)/5$ and $Z^2 = (4 - Z)/3$, then Z^1 and Z^2 are again nonnegative elements of $\mathscr{M}(P)$ satisfying (8), $E(Z^1_\infty) = E(Z^2_\infty) = 1$ and $E(Z^1_0 | \mathscr{F}^0_0) = E(Z^2_0 | \mathscr{F}^0_0) = 1$. Moreover if $P^1 = Z^1_\infty \cdot P$ and $P^2 = Z^2_\infty \cdot P$, then $P^1 \neq P \neq P^2$ and $P = (5P^1 + 3P^2)/8$. But Lemma 3 implies that P^1 and P^2 belong to $\mathscr{S}(\mathscr{Q})$, which contradicts extremality for P.

Now we prove part (ii). We suppose that $\mathscr{Q} = \{\boldsymbol{Q}\}$, where $\boldsymbol{Q} = (Q, \nu, W, A, B)$. Let $P', P'' \in \mathscr{S}(\mathscr{Q})$, and $P = aP' + bP''$ be a convex combination of P' and P''. It is evident that the restriction of P to $(\Omega, \mathscr{F}^0_0)$ is Q. As $M^P_\mu = aM^{P'}_\mu + bM^{P''}_\mu$ and $M^P_\nu = aM^{P'}_\nu + bM^{P''}_\nu$, it is easy to see that M^P_μ is $\tilde{\mathscr{P}}$-σ-finite and that ν is the dual predictable projection of μ for P. As A^i and $C(\nu, W^i)$ (this is the process associated to W^i and ν in part (b) of the proof of Lemma 3) are in $\mathscr{V}_{\text{loc}}(P') \cap \mathscr{V}_{\text{loc}}(P'')$ and are predictable, they also belong to $\mathscr{V}_{\text{loc}}(P)$ (because $\mathscr{A}(P') \cap \mathscr{A}(P'') = \mathscr{A}(P)$).

It is evident that $\mathscr{M}(P') \cap \mathscr{M}(P'') \subset \mathscr{M}(P)$. Now if $M \in \mathscr{M}_{\text{loc}}(P') \cap \mathscr{M}_{\text{loc}}(P'')$ is continuous, let $T_n = \inf(t\colon M_t \geqq n)$: We have $M^{T_n} \in \mathscr{M}(P') \cap \mathscr{M}(P'')$ and T_n tends towards infinity both P'-a.s. and P''-a.s. Therefore $M^{T_n} \in \mathscr{M}(P)$ and T_n tends towards infinity P-a.s.; thus $M \in \mathscr{M}_{\text{loc}}(P)$. Applying this to processes M^i and $M^iM^j - B^{ij}$, we see that these processes are in $\mathscr{M}_{\text{loc}}(P)$. Thus $P \in \mathscr{S}(\mathscr{Q})$, and this set is convex.

Let us assume now that $P \in \mathscr{S}(\mathscr{Q})$ is not extremal in $\mathscr{S}(\mathscr{Q})$, that is there exist P' and P'' in $\mathscr{S}(\mathscr{Q})$ and a number $0 < a < 1$ such that $P = aP' + (1 - a)P''$ and that $P' \neq P''$. Therefore $P' \ll P$ and $P' \neq P$.

Let $Z_\infty = dP'/dP$ and $Z_t = E(Z_\infty | \mathscr{F}_t)$. Then $Z = (Z_t)$ is an element of $\mathscr{M}(P)$ which satisfies (8) from Lemma 3 (because $\mathscr{Q} = \{\boldsymbol{Q}\}$). Let us assume that condition (b) holds true: Then $\langle Z^c, M^i \rangle = 0$ for each $i \in I$ implies $Z^c = 0$, and using Theorem 1 and (6) we see that $Z^d_t = Z_0$ P-a.s. for each t. But the restrictions of P and P' to $(\Omega, \mathscr{F}^0_0)$ are the same, which implies $E(Z_0 | \mathscr{F}^0_0) = 0$; as $P' \neq P$ we do not have $Z_0 = 1$ P-a.s. and therefore Z_0 is not P-a.s. equal to $E(Z_0 | \mathscr{F}^0_0)$: In other words, condition (a) fails, and part (ii) is proven.

4. Examples and applications.

1. *Some examples for continuous processes.* In many martingale problems one considers only continuous processes. That is, one starts with a family $X = (X^i)_{i \in I}$ (where I is usually finite) of continuous adapted processes, and a set $\mathscr{Q}^c$ consisting of triplets $\boldsymbol{Q}^c = (Q, A, B)$ where

Q is a probability measure on $(\Omega, \mathscr{F}^0_0)$,

$A = (A^i)_{i \in I}$ and $B = (B^{ij})_{i,j \in I}$ are two families of continuous adapted processes.

DEFINITION. *A probability measure* P on $(\Omega, \mathscr{F})$ is a *solution of the martingale problem* $(X, \mathscr{F}^0_0, \mathscr{Q}^c)$ if there exists $\boldsymbol{Q}^c = (Q, A, B) \in \mathscr{Q}^c$ such that:

(i) Q is the restriction of P to $(\Omega, \mathscr{F}^0_0)$.

(ii) For each $i \in I$, $A^i \in \mathscr{V}_{\text{loc}}(P)$ and $M_t^i = X_t^i - X_0^i - A_t^i$ defines a (continuous) element of $\mathscr{M}^2_{\text{loc}}(P)$.

(iii) For each $i, j \in I$ we have $\langle M^i, M^j \rangle = B^{ij}$.

We denote by $\mathscr{S}^c(\mathscr{Q}^c)$ the set of all solutions of the martingale problem $(X, \mathscr{F}_0^0, \mathscr{Q}^c)$. *Such a problem is a particular case of the general problem* examined in §§2 and 3: Actually if we put $\mu = 0$ and if $\mathscr{Q}$ is the set of all those $\boldsymbol{Q} = (Q, \nu, W, A, B)$ such that $\nu = 0$, $W^i = 0$ for each $i \in I$, and $(Q, A, B) \in \mathscr{Q}^c$, then we have $\mathscr{S}^c(\mathscr{Q}^c) = \mathscr{S}(\mathscr{Q})$.

Below we give some examples of martingale problems $(X, \mathscr{F}_0^0, \mathscr{Q}^c)$ and applications of Theorem 2. In all these examples, *$(\mathscr{F}_t)$ is the smallest increasing right-continuous family of σ-algebras for which all X^i are adapted and $\mathscr{F}_0^0 = \sigma(X_0^i, i \in I)$.* But before giving these examples, we end this introduction by a very important remark.

REMARK. Let $\hat{\Omega}$ be the set of all continuous functions from $[0, \infty[$ into $\boldsymbol{R}^I$, with the "canonical" process and σ-algebras $\hat{X}$ and $(\hat{\mathscr{F}}_t)$. As $(\mathscr{F}_t)$ is "generated" by the process X, it is easily seen that any martingale problem $(X, \mathscr{F}_0^0, \mathscr{Q}^c)$ can be transported onto $\hat{\Omega}$ (by the application $\Omega \ni \omega \rightsquigarrow \hat{\omega} = X\cdot(\omega) \in \hat{\Omega}$) as a martingale problem $(\hat{X}, \hat{\mathscr{F}}_0^0, \hat{\mathscr{Q}}^c)$. Moreover if $P \in \mathscr{S}^c(\mathscr{Q}^c)$, the distribution of X under P is a probability measure $\hat{P}$ on $(\hat{\Omega}, \hat{\mathscr{F}})$ belonging to $\mathscr{S}^c(\hat{\mathscr{Q}}^c)$, and $P \neq P'$ implies $\hat{P} \neq \hat{P}'$ (because $\mathscr{F} = \sigma(X_t^i, i \in I, t \geqq 0)$, of course).

In the literature, martingale problems are very often stated with $\Omega = \hat{\Omega}$, and we obtain conditions on $\mathscr{Q}^c$ for both existence and uniqueness of the solution. Here Ω is an arbitrary space: Therefore in the following examples we do not give conditions on $\mathscr{Q}^c$ for the existence of a solution (and actually when Ω is "too small" no solution exists). But *when a solution exists, it is unique whenever we have uniqueness for the transported problem $(\hat{X}, \hat{\mathscr{F}}_0^0, \hat{\mathscr{Q}}^c)$ on the canonical space.*

(a) *Wiener process.* We suppose that X consists in only one process (X_t) and that $\mathscr{Q}^c = \{\boldsymbol{Q}^c\}$, where $\boldsymbol{Q}^c = (Q, A, B)$ is defined by

Q gives mass 1 to the atom $\{X_0 = 0\}$ of $\mathscr{F}_0^0$,

$A_t = 0$ and $B_t = t$.

P. Lévy has shown that when this martingale problem admits a solution P, then under P the process X is a Wiener process, thus implying uniqueness for this solution. Then Theorem 2 asserts that any element of $\mathscr{M}_{\text{loc}}(P)$ is a stochastic integral with respect to X, and that the "0-1 law" holds for $\mathscr{F}_0$: These are of course classical results (cf. for example Motoo and Watanabe [**20**]), and various proofs of them have been given. Let us just say that a proof using a (simple) version of Theorem 2 is due to Dellacherie [**4**].

(b) *Diffusion processes.* We suppose that $I = \{1, 2, \cdots, d\}$ and we put $X_t = (X_t^1, \cdots, X_t^d)$, which defines a $\boldsymbol{R}^d$-valued process (X_t). Let $\mathscr{Q}_x^c = (Q_x, A, B)$ be defined by

Q_x gives mass 1 to the atom $\bigcap_{i=1}^d \{X_0^i = x^i\}$, where $x = (x^i) \in \boldsymbol{R}^d$,

$A_t^i = \int_0^t a^i(X_s)\, ds$,

$B_t^{ij} = \int_0^t b^{ij}(X_s)\, ds$,

a^i and b^{ij} being real-valued Borel functions on $\boldsymbol{R}^d$, with $b^{ij} = b^{ji}$ and $\sum y_i y_j b^{ij} \geqq 0$ ident cally for any $y_i \in \boldsymbol{R}$.

Under various conditions on coefficients a^i and b^{ij} (for example a^i is bounded, b^{ij} is bounded and continuous, and for any $(y_i) \in \boldsymbol{R}^d \backslash \{0\}$ we have identically $\sum y_i y_j b^{ij} > 0$: cf. Stroock and Varadhan [**22**]), $\mathscr{S}^c(\mathscr{Q}_x^c)$ admits at most one point

P_x (when this unique P_x exists for each x, then under P_x, X is a diffusion process starting at point x, with diffusion coefficients b^{ij} and drift coefficients a^i).

Then *if* $\mathscr{S}^c(\mathscr{Q}_x^c)$ *contains exactly one point* P_x, *each element* M *of* $\mathscr{M}_{\text{loc}}(P_x)$ *can be written as*

(11) $$M_t = E_x(M_0) + \sum_{i=1}^{d} u^i \cdot (X^i - A^i)$$

for suitable $u^i \in L^2_{\text{loc}}(X^i - A^i, P)$: This comes from the remark following Proposition 2, after noticing that the hitting times of the sets

$$\left\{x: \inf_{(y_i)\in \boldsymbol{R}^d\{0\}} \frac{1}{\|y\|^2} \sum y_i y_j b^{ij}(x) < \frac{1}{n}\right\}$$

increase P_x-a.s. to infinity. For similar results under different assumptions, cf. Motoo and Watanabe **[20]**.

(c) *Generalized diffusion processes.* X is as in example (b), and $\mathscr{Q}^c = \{(Q, A, B)\}$ is defined by

Q gives mass 1 to the atom $\bigcap_{i=1}^d \{X_0^i = x^i\}$, where $x = (x^i) \in \boldsymbol{R}^d$,

$A_t^i = \int_0^t \tilde{a}_s^i\, ds$,

$B_t^{ij} = \int_0^t \tilde{b}_s^{ij}\, ds$,

$\tilde{a}^i$ and $\tilde{b}^{ij}$ being well-measurable processes with $\tilde{b}_t^{ij} = \tilde{b}_t^{ji}$ and $\sum y_i y_j \tilde{b}_t^{ij} \geqq 0$ identically for any $y_i \in \boldsymbol{R}$.

When $P \in \mathscr{S}^c(\mathscr{Q}^c)$ we say that under P, *the process* (X_t) *is a generalized diffusion process* (Lipčer and Shyriaev **[16]**). If we suppose for example that $\tilde{a}^i$ is bounded and that $\tilde{b}_t^{ij} = b^{ij}(X_t)$ where b^{ij}'s satisfy Stroock and Varadhan's condition of the example above, then $\mathscr{S}^c(\mathscr{Q}^c)$ admits at most one solution **[16]**. Then under this condition, when $P \in \mathscr{S}^c(\mathscr{Q}^c)$ any $M \in \mathscr{M}_{\text{loc}}(P)$ can be written as (11): This extends the representation results given by Fujisaki, Kallianpur and Kunita **[8]** and may be used in filtering theory.

(d) *Modulus of Wiener process.* We suppose that X consists in only one process (X_t) which satisfies $X_t \geqq 0$ and $\int 1_{\{0\}}(X_s)\, ds = 0$ identically. $\mathscr{Q}^c$ contains all those triplets $\boldsymbol{Q}^c = (Q, A, B)$ such that

Q gives mass 1 to the atom $\{X_0 = 0\}$,

$A = (A_t)$ is any continuous adapted increasing process satisfying identically

(12) $$A_t = \int_0^t 1_{\{0\}}(X_s)\, dA_s,$$

$B_t = t$.

Watanabe **[25]** has shown that $\mathscr{S}^c(\mathscr{Q}^c)$ *contains at most one point*, and if P is the only one element of $\mathscr{S}^c(\mathscr{Q}^c)$ then under P, *the process* (X_t) "*is*" *the modulus of a Wiener process* (i.e., has the same distribution as the modulus of a Wiener process). In addition if $\boldsymbol{Q}^c = (Q, A, B)$ is associated to P, then A is the so-called *local time* of the set $\{0\}$.

Therefore *if* $P \in \mathscr{S}^c(\mathscr{Q}^c)$ (or in other words: *if* (X_t) *is the modulus of a Wiener process under P*) *then any* $M \in \mathscr{M}_{\text{loc}}(P)$ *is*

$$M_t = E(M_0) + u \cdot (X - A)$$

where A is the (unique) continuous increasing process satisfying (12) and $X - A \in$

$\mathcal{M}_{\text{loc}}(P)$, and where $u \in L^2_{\text{loc}}(X - A, P)$; in addition, the local martingale $X - A$ is itself a Wiener process (this result is due again to Motoo and Watanabe **[20]**, as well as some particular cases of the next example).

REMARK. In order to set this martingale problem in a proper form, we have followed El-Karoui [7]. However in [7] condition "$\int 1_{\{0\}}(X_s)\, ds = 0$ holds identically" is dropped, while for P to be a solution one adds the condition "$\int 1_{\{0\}}(X_s)\, ds = 0$ P-a.s.": It is evident that these two martingale problems are equivalent, as far as uniqueness is concerned. The same remark will also hold for the next example.

(e) *Diffusion processes with reflection.* We still follow El-Karoui [7]. We suppose that $I = \{1, 2, \cdots, d\}$ and that $X_t = (X_t^1, \cdots, X_t^d)$ is a process with values in $\bar{G} = [0, \infty[\times \boldsymbol{R}^{d-1}$ such that $\int 1_{\partial G}(X_t)\, dt = 0$ identically, where $\partial G = \{0\} \times \boldsymbol{R}^{d-1}$. $\mathcal{Q}_x^c$ contains all the triplets (Q_x, A, B) such that

Q_x gives mass 1 to the atom $\bigcap_{i=1}^d \{X_0^i = x^i\}$, where $x = (x^i) \in \bar{G}$,

$C = (C_t)$ is any continuous adapted increasing process satisfying identically

$$C_t = \int_0^t 1_{\partial G}(X_s)\, dC_s,$$

and

$$A_t^i = \int_0^t a^i(X_s)\, ds + \int_0^t \gamma^i(X_s)\, dC_s, \qquad B_t^i = \int_0^t b^{ij}(X_s)\, ds + \int_0^t \alpha^{ij}(X_s)\, dC_s.$$

In these formulas, a^i, b^{ij}, γ^i and α^{ij} are Borel functions on $\bar{G}$: b^{ij} is like in example (b), $\alpha^{ij} = \alpha^{ji}$ and $\sum y_i y_j \alpha^{ij} > 0$ for any $y_i \in \boldsymbol{R}$, $\alpha^{1i} = 0$ for any $i \leqq d$, and $\gamma^1 = 1$ identically.

If $\mathcal{S}^c(\mathcal{Q}_x^c)$ contains exactly one point P_x for each $x \in \bar{G}$, then under P_x, (X_t) *is a diffusion process in* $\bar{G}$, *starting at* x, *with a reflecting boundary* ∂G; inside $\bar{G} - \partial G$ the motion is described by a^i and b^{ij}, while $\gamma^1 = 1$ means that the process does not stay on ∂G for a positive time, and γ^i and α^{ij} $(i, j \geqq 2)$ describe how the particle moves along ∂G.

Then *when* P_x *is the unique solution of* $\mathcal{S}^c(\mathcal{Q}_x^c)$, *any* $M \in \mathcal{M}_{\text{loc}}(P_x)$ *can be written as* (11): For example this holds when a^i, γ^i and the elements of the square roots of matrices (b^{ij}) and (α^{ij}) are bounded and Lipschitz ([7] ,Watanabe **[25]**).

2. *Some examples for discontinuous processes.* Now we come back to the general situation described in §2.

(a) *Point processes.* A *multivariate point process* is an integer-valued random measure μ which satisfies

$$\mu(dt, dx) = \sum_{(n)} 1_{\{T_n < \infty\}}\, \varepsilon_{(T_n, X_n)}(dt, dx),$$

where (T_n) is a *strictly increasing* sequence of stopping times and X_n is an $\mathcal{F}_{T_n}$-measurable random variable with values in E. When E reduces to one point, μ is completely described by the counting process $N_t = \mu(]0, t] \times E)$ and we say that we have a *point process.*

Let μ be a multivariate point process, and $I = \varnothing$. We suppose that $(\mathcal{F}_t)$ is the smallest increasing right-continuous family of σ-algebras for which μ is well-measurable, and $\mathcal{F}_0^0 = \{\Omega, \varnothing\}$. Let $\mathcal{Q} = \{\boldsymbol{Q}\}$ where $\boldsymbol{Q} = (Q, \nu, W, A, B)$ is completely described by ν (because Q, W, A and B are necessarily trivial). Then one knows [12] that $\mathscr{S}(\mathscr{Q})$ *contains at most one point P: Therefore if P is any probability measure on $(\Omega, \mathscr{F})$, each $M \in \mathscr{M}_{\text{loc}}(P)$ can be written as*

$$M_t = E(M_0) + U*(\mu - \nu)_t$$

where $U \in \mathscr{G}_{\text{loc}}(\mu, P)$, and ν is the dual predictable projection of μ for P. This representation result was shown in [12] (and by Dellacherie [4] for Poisson processes) by the same method, and also by Chou and Meyer [1] (for point processes) and Davis [2] by a different way.

(b) *Multidimensional semimartingales.* All our previous examples (except the case of multivariate point processes) fall into the following framework, which is very common in practice. We suppose I is finite or denumerable, we put $E = \boldsymbol{R}^I$ (we denote by $x = (x^i)$ a point of E), and we start with a process $(Y_t) = (Y_t^i, i \in I)$ with values in E, whose each coordinate Y^i is adapted, right-continuous and left-hand limited.

We are interested in probability measures P on $(\Omega, \mathscr{F})$ *for which* (Y_t) *is a multidimensional semimartingale* (i.e., each Y^i is a semimartingale; recall from [6] that X is a semimartingale if one can write $X_t = X_0 + A_t + M_t$, where $M \in \mathscr{M}_{\text{loc}}(P)$ and $A \in \mathscr{A}(P)$). It turns out that such P are solutions of martingale problems in the sense of §2.

More precisely define μ by (2) (where $\Delta Y_t = (\Delta Y_t^i) \in E$) and put

$$X_t^i = Y_t^i - \sum_{s \leq t} \Delta Y_s^i 1_{\{|\Delta Y_s^i| > 1\}}.$$

It is evident that for any solution P of any martingale problem associated to (μ, X), then (Y_t) is a multidimensional semimartingale.

Conversely let P be a probability measure for which (Y_t) is a multidimensional semimartingale: ν denotes the dual predictable projection of μ for P; as X^i has only bounded jumps ($|\Delta X^i| \leq 1$) it has a unique decomposition $X^i = X_0^i + A^i + N^i$ where $N^i \in \mathscr{M}^2_{\text{loc}}(P)$ and A^i is a predictable element of $\mathscr{V}_{\text{loc}}(P)$ (cf. [14], or Meyer [19]); at last let $B^{ij} = \langle (N^i)^c, (N^j)^c \rangle$ and $U^i(\omega, t, (x^j)) = x^i 1_{\{|x^i| \leq 1\}}$, which is a $\tilde{\mathscr{P}}$-measurable function on $\tilde{\Omega}$.

PROPOSITION 4. *Let μ and X be as before. Let P be a probability measure on $(\Omega, \mathscr{F})$ for which (Y_t) is a multidimensional semimartingale and whose restriction to $(\Omega, \mathscr{F}_0^0)$ is Q. Then with the above notations we have $P \in \mathscr{S}(\{\boldsymbol{Q}\})$, where $\boldsymbol{Q} = (Q, \nu, (U^i), (A^i), (B^{ij}))$.*

PROOF. μ being associated to (Y_t) by (2), we know that M_μ^P is $\tilde{\mathscr{P}}$-σ-finite [13]. The only things to show are that $U^i \in \mathscr{G}^2_{\text{loc}}(\mu, P)$ and that $M^i = N^i - U^i * (\mu - \nu)$ is continuous for each $i \in I$.

Let us recall a result from [13]: If V is a $\tilde{\mathscr{P}}$-measurable function on $\tilde{\Omega}$ and if T is a predictable stopping time, then

$$\int_E \nu(\{T\}, dx) V(T, x) = E\left[\int_E \mu(\{T\}, dx) V(T, x) \,\middle|\, \mathscr{F}_{T-}\right] \tag{13}$$

on $\{T < \infty\}$ whenever these integrals make sense.

We have $\Delta N_t^i = U^i(t, \Delta Y_t) - \Delta A_t^i$, which implies that $M_\mu^P(\Delta N^i | \tilde{\mathscr{P}}) = U^i - \Delta A^i$

by the definition of μ, and also that for any predictable stopping time T such that ΔA_T^i and ΔN_T^i are integrable,

$$\int_E \nu(\{T\}, dx) U^i(T, x) - \Delta A_T^i = E(\Delta N_T^i \mid \mathscr{F}_{T-}) = 0$$

(apply (13)). Therefore the section theorem for predictable sets **[3]** yields

$$\Delta A_t^i = \int_E \nu(\{t\}, dx) U^i(t, x) \tag{14}$$

except on a P-null set (not depending on t). Thus if we put (for a fixed i) $U = U^i - \Delta A^i$ and if we compute W by (3), we get $W = U^i$. Then Theorem 1 implies $U^i \in \mathscr{G}^2_{\mathrm{loc}}(\mu, P)$. At last an easy computation, using (14), shows that

$$\Delta[U^i * (\mu - \nu)]_t = U^i(t, \Delta Y_t) - \Delta A_t^i = \Delta N_t^i,$$

and $N^i - U^i * (\mu - \nu)$ is continuous. □

For one-dimensional semimartingales, the above problem has been studied in detail in **[14]**; the family (ν, A, B) is called the set of *local characteristics* of (Y_t) (cf. Grigelionis **[10]**). As U^i is independent of P, this family together with Q are sufficient to determine completely the martingale problem.

(c) *Processes with independent increments.* We stay in the previous framework, with $I = \{1, 2, \cdots, d\}$. In addition we assume that $(\mathscr{F}_t)$ is the smallest increasing right-continuous family of σ-algebras for which (Y_t) is adapted, and $\mathscr{F}_0^0 = \sigma(Y_0)$. We suppose that $\mathscr{Q} = \{(Q, \nu, (U^i), A, B)\}$, where

Q gives mass 1 to the atom $\bigcap_{i=1}^d \{Y_0^i = 0\}$,

$\nu(\omega; dt, dx) = dt\ F(dx)$, where F is a positive measure on $\boldsymbol{R}^d$ such that $\int(|x|^2 \wedge 1)\ F(dx)$ is finite,

$A_t^i = a^i t$ and $B_t^{ij} = b^{ij} t$, where (a^i) is a set of real numbers and (b^{ij}) is a symmetric matrix.

It is well known that if $P \in \mathscr{S}(\mathscr{Q})$ does exist, then under P, the process (Y_t) is a *process with independent increments with Lévy measure F, drift a^i and diffusion coefficients b^{ij}*: Therefore P is unique.

In particular Theorem 2 implies that *any $M \in \mathscr{M}_{\mathrm{loc}}(P)$ is*

$$M_t = E(M_0) + V * (\mu - \nu) + \sum_{i=1}^d u^i \cdot M^i \tag{15}$$

where $V \in \mathscr{G}_{\mathrm{loc}}(\mu, P)$, $M^i = X^i - U^i * (\mu - \nu)$ and $u^i \in L^2_{\mathrm{loc}}(M^i, P)$ (here, M^i is a rescaled Wiener process; the above formula follows easily from the remark after Proposition 2, because B^{ij}'s are deterministic). Moreover if $M \in \mathscr{M}^2_{\mathrm{loc}}(P)$, then $V \in \mathscr{G}^2_{\mathrm{loc}}(\mu, P)$. This representation result has been shown by Grigelionis **[11]** and Galtčuk **[9]** in case $M \in \mathscr{M}^2(P)$, and also in **[14]** in case $d = 1$.

(d) *Diffusion processes with jumps.* We make the same assumptions as in example (c). Let $\mathscr{Q} = \{(Q_y, \nu, (U^i), A, B)$, where

Q_y gives mass 1 to the atom $\bigcap_{i=1}^d \{Y_0^i = y^i\}$, where $y = (y^i) \in \boldsymbol{R}^d$,

$\nu(\omega; dt, ds) = dt N(Y_{t-}(\omega), dx)$,

$A_t^i = \int_0^t a^i(Y_s)\, ds$,

$B_t^{ij} = \int_0^t b^{ij}(Y_s)\, ds$,

where a^i and b^{ij} are as in the continuous diffusion case, and N is a positive kernel on $\boldsymbol{R}^d$ such that $N(x, \{0\}) = 0$.

This formulation is slightly different from that of Stroock [**23**], the difference amounting to the fact that the coefficients a^i are not what is called drift coefficient in [**23**] (we let the reader find the details himself). But the results of [**23**] still hold: For example if a^i and b^{ij} satisfy Stroock and Varadhan's condition (example 4.1(b)) and if $\int_E f(y)\,(y^2 \wedge 1)N(x, dy)$ is bounded and continuous for each bounded continuous function f, then (Q_y) *admits at most one point* P_y (when this P_y exists for each $y \in \boldsymbol{R}^d$, then under P_y, Y is a (Markov) "diffusion process with Lévy generator", the Lévy kernel of this Markov process being $N(x, dy - x)$).

Then if $\mathscr{S}(Q_y)$ contains exactly one point P_y, *any* $M \in \mathscr{M}_{\text{loc}}(P_y)$ *can be written as* (15).

References

1. C. S. Chou and P.-A. Meyer, *La représentation des martingales relatives à un processus ponctuel discret*, C. R. Acad. Sci. Paris Sér. A-B **278** (1974), A1561–A1563.

2. M. H. A. Davis, *The representation of martingales of jump processes*, SIAM J. Control and Optimization **14** (1976), 623–638.

3. C. Dellacherie, *Capacités et processus stochastiques*, Springer-Verlag, Berlin and New York, 1972.

4. ———, *Intégrales stochastiques par rapport aux processus de Wiener ou de Poisson*, Séminaire de Probabilités. VIII (Univ. Strasbourg, 1972–1973), Lecture Notes in Math., vol. 381, Springer-Verlag, Berlin and New York, 1974, pp. 25–26; correction, Séminaire de Probabilités. IX, Lecture Notes in Math., vol. 465, Springer-Verlag, Berlin and New York, 1975, p. 494. MR **51** #6981a,b.

5. C. Doléans-Dade, *Quelques applications de la formule de changement de variables pour les semimartingales*, Z. Wahrscheinlichkeitstheorie und Verw. Gebiete **16** (1970), 181–194. MR **44** #1113.

6. C. Doléans-Dade and P.-A. Meyer, *Intégrales stochastiques par rapport aux martingales locales*, Séminaire de Probabilités. IV (Univ. Strasbourg, 1968/69), Lecture Notes in Math., vol. 124, Springer-Verlag, Berlin and New York, 1970, pp. 77–107. MR **42** #5313.

7. N. E. Karoui, *Processus de réflexion dans* $\boldsymbol{R}^n$, Séminaire de Probabilités. IX (Univ. Strasbourg, 1973/74), Lecture Notes in Math., vol. 465, Springer-Verlag, Berlin and New York, 1975.

8. M. Fujisaki, G. Kallianpur and H. Kunita, *Stochastic differential equations for the non-linear filtering problem*, Osaka J. Math. **9** (1972), 19–40. MR **49** #1574.

9. L. Galtčuk, *The structure of a class of martingales*, Proc. School-Seminar on Random Processes (Vilnius), Acad. Sci. Lit. SSR **I** (1975), 7–32. (Russian)

10. B. Grigelionis, *On non-linear filtering theory and absolute continuity of measures, corresponding to stochastic processes*, Second Japan-USSR Sympos., Lecture Notes in Math., vol. 330, Springer-Verlag, Berlin and New York, 1973.

11. ———, *On the stochastic integral representation of square-integrable martingales*, Lit. Math. J. **14** (1974), no. 4, 53–69. (Russian)

12. J. Jacod, *Multivariate point processes: predictable projection, Radon-Nikodym derivatives, representation of martingales*, Z. Wahrscheinlichkeitstheorie und Verw. Gebiete **31** (1975), 235–253.

13. ———, *Un théorème de répresentation pour les martingales discontinues*, Z. Wahrscheinlichkeitstheorie und Verw. Gebiete **34** (1976), 225–244.

14. J. Jacod and J. Memin, *Caractéristiques locales et conditions de continuité absolue pour les semimartingales*, Z. Wahrscheinlichkeitstheorie und Verw. Gebiete **35** (1976), 1–37.

15. Ju. M. Kabanov, *Representation of functionals of Wiener and Poisson processes in the form of stochastic integrals*, Teor. Verojatnost. i Primenen **18** (1973), 376–380 = SIAM Theor. Probability Appl. **18** (1973), no. 2, 362–365. MR **47** #7809.

16. R. Š. Lipčer and A. Shyriaev, *Statistiques des processus stochastiques*, "Nauka", Moscow, 1974. (Russian)

17. P.-A. Meyer, *Probabilités et potentiel*, Publ. Inst. Math. Univ. Strasbourg, no. 14, Actualités Sci. Indust., no. 1318, Hermann, Paris, 1966. MR **34** #5118.

18. P.-A. Meyer, *Intégrales stochastiques*, I, II, III, IV, Séminaire de Probabilités (Univ. Strasbourg, Strasbourg, 1966/67), Vol. I, Lecture Notes in Math., vol. 39, Springer-Verlag, Berlin, 1967, pp. 72–94, 95–117, 118–141, 142–162. MR **37** #7000.

19. ———, *Un cours sur les integrales stochastiques* Séminaire de Probabilités. X (Univ. Strasbourg, 1974/75), Lecture Notes in Math., vol. 511, Springer-Verlag, Berlin and New York, 1976.

20. M. Motoo and S. Watanabe, *On a class of additive functionals of Markov processes*, J. Math. Kyoto Univ. **4** (1964/65), 429–469. MR **33** #4994.

21. A. V. Skorohod, *Studies in the theory of random processes*, Izdat. Kiev. Univ., Kiev, 1961; English transl., Addison-Wesley, Reading, Mass., 1965. MR **32** #3082 a,b.

22. D. W. Stroock and S. R. S. Varadhan, *Diffusion processes with continuous coefficients*. I, II, Comm. Pure Appl. Math. **22** (1969), 345–400; 479–530. MR **40** #6641; 8130.

23. D. W. Stroock, *Diffusion processes associated with Lévy generators*, Z. Wahrscheinlichkeitstheorie und Verw. Gebiete **32** (1975), 209–244.

24. J. H. Van Schuppen and E. Wong, *Transformation of local martingales under a change of law*, Ann. Probability **2** (1974), 879–888. MR **50** #11426.

25. S. Watanabe, *On stochastic differential equations for multi-dimensional diffusion processes with boundary conditions*, J. Math. Kyoto Univ. **11** (1971), 169–180. MR **43** #1291.

UNIVERSITE DE RENNES

Proceedings of Symposia in Pure Mathematics
Volume 31, 1977

CENTRAL LIMIT THEOREM AND RELATED QUESTIONS IN BANACH SPACE

NARESH C. JAIN*

Introduction. The central limit problem for independent summands taking values in a separable infinite dimensional Banach space was first considered by Fortet and Mourier **[6]**, **[7]**, **[25]**. In the case of a separable Hilbert space a complete development is given by Varadhan **[34]**; this work may also be found in **[28]**. A similar formulation for triangular arrays can be given for spaces of type 2 and cotype 2 (definitions given below) as is done in Garling **[8]**. We will discuss mainly the case of independent identical summands where definitive answers to certain questions are now known.

In addition to the original papers referenced here, the two survey paper **[15]**, **[31]**, which also contain some new results, give a good deal of information on the subject we are going to discuss. However, since these papers were written some more work has been done which will be included here.

In what follows B will denote a real separable Banach space and B^* its dual (bounded linear functionals on B). θ will denote the null element of B. $\mathscr{B}$ is the σ-algebra of Borel subsets of B. A mapping X from a probability space $(\Omega, \mathscr{F}, P)$ into B is a B-valued random variable (r.v.) if $f(X)$ is a real-valued r.v. on $(\Omega, \mathscr{F}, P)$ for every $f \in B^*$. The probability measure μ_X induced on $(B, \mathscr{B})$ by a B-valued r.v. X is called its distribution. Let M denote the class of all probability measures on $(B, \mathscr{B})$. For $p \geqq 1$ we will consider.

$$WM_0^p = \left\{\mu \in M: \int |f|^p \, d\mu < \infty, \int f \, d\mu = 0\right\},$$

AMS (*MOS*) *subject classifications* (1970). Primary 60B10, 60F05.

Key words and phrases. Central limit theorem, Banach space valued random variables, reproducing kernel Hilbert space, spaces of type 2 and cotype 2.

*Supported partially by the National Science Foundation.

$$M_0^p = \left\{\mu \in M: \int \|x\|^p \, d\mu(x) < \infty, \int x \, d\mu(x) = \theta\right\},$$

$$G = \{\mu: \mu \in WM_0^2, f \text{ Gaussian on } (B, \mathscr{B}, \mu), \forall f \in B^*\};$$

thus G is the class of all Gaussian measures. If $\mu \in WM_0^2$, its covariance kernel Γ_μ is given by $\Gamma_\mu(f, g) = \int fg \, d\mu, f, g \in B^*$. We then define the class of *pregaussian* measures by

$$\text{PG} = \{\mu: \mu \in WM_0^2, \Gamma_\mu = \Gamma_\nu \text{ for some } \nu \in G\}.$$

For convenience we will also say that a B-valued r.v. belongs to one of these classes if its distribution does. We write $\Gamma_\mu = \Gamma_X$ if $\mu = \mu_X$.

If μ_n, $\mu \in M$, $\mu_n \xrightarrow{w} \mu$ denotes the weak convergence of probability measures, i.e., $\int f \, d\mu_n \to \int f \, d\mu$ for every bounded continuous function on B.

Let X_1 be a B-valued r.v.; then S_n, $n \geqq 1$, denotes $X_1 + X_2 + \cdots + X_n$, where $\{X_j, j \geqq 1\}$ is a sequence of independent, identically distributed, B-valued r.v.'s. The class of all random variables that satisfy the central limit theorem is denoted by

$$\text{CLT} = \{X_1: \mu_{X_1} \in WM_0^2, \mu_{S_n/\sqrt{n}} \xrightarrow{w} \text{some } \nu \in M\}.$$

It is clear by the convergence of finite dimensional distributions that if $X_1 \in \text{CLT}$ and ν is the limit measure, then $\nu \in G$ and $\Gamma_{X_1} = \Gamma_\nu$. Therefore

$$\text{CLT} \subset \text{PG}. \tag{1.1}$$

If B is isomorphic to a separable Hilbert space H, then it easily follows from Varadhan's results **[34]** and also from the results given below that CLT = PG = M_0^2. One may ask for what Banach spaces does one have CLT = PG or CLT = M_0^2? It was known from Vakhania's results **[32]** that CLT $\neq$ PG even for $B = l^p$, $p > 2$. More recently Dudley **[4]** showed that no moment conditions suffice; in fact, he gives an example in $C[0,1]$ where $X \in \text{PG}$, $\|X\| \leqq 1$ a.s. $\not\Rightarrow$ $X \in \text{CLT}$. Before describing the main results it is necessary to define Banach spaces of type 2 and cotype 2.

A Banach space B is of *type* 2 (*cotype* 2) if there exists $c < \infty$ such that given $X_1, X_2, \cdots, X_n \in M_0^2$, independent r.v.'s,

$$E(\|X_1 + \cdots + X_n\|^2) \underset{(\geqq)}{\leqq} c \sum_{j=1}^n E(\|X_j\|^2).$$

Examples of spaces of type 2 are the L^p spaces of a σ-finite measure space, $2 \leqq p < \infty$; a simple direct proof of this result (known earlier) may be found in **[15]**. If $1 \leqq p \leqq 2$, these spaces are of cotype 2 **[26]**.

The following theorem of Kawapien **[22]** will be used.

THEOREM 1. *B is of type 2 and cotype 2 $\Leftrightarrow$ B is isomorphic to a Hilbert space H.*

Some of the main results on the central limit theorem that have been recently discovered are the following.

THEOREM 2 (HOFFMANN-JØRGENSEN AND PISIER **[13]**). *$M_0^2 \subset CLT \Leftrightarrow B$ is of type 2.*

THEOREM 3. *$CLT \subset M_0^2 \Leftrightarrow B$ is of cotype* 2.

By Theorem 1 we get the following obvious corollary of Theorems 2 and 3.

THEOREM 4. *$CLT = M_0^2\ (=PG) \Leftrightarrow B$ is isomorphic to a Hilbert space.*

Another natural question is answered by

THEOREM 5. *$CLT = PG \Leftrightarrow B$ is of cotype* 2.

The main ingredients of the proof of Theorem 5 were there in a paper of Maurey **[26]** as observed by Pisier **[31]**. N. N. Vakhania informed us that Chobanian and Tarieladze also proved it essentially along the same lines independently. Their work has not appeared in print as yet.

The proofs of Theorems 2, 3 and 5 are discussed in §4. In §2 we discuss the reproducing kernel Hilbert space (rkhs) of a measure in WM_0^2 which is of some intrinsic interest and is useful (but not indispensable) to prove Theorem 5. §3 gives a useful result which says that $X \in \mathrm{CLT} \Rightarrow E(\|X\|^{2-\varepsilon}) < \infty$ for $\varepsilon > 0$.

One important question is to characterize those r.v.'s which belong to CLT for a given Banach space (characterization of the domain of normal attraction of a Gaussian measure). Such results are discussed in §5. Some remarks are made in §6.

It is sometimes helpful to regard B as a closed subspace of $C[0, 1]$ (Banach-Mazur). If $x \in C[0, 1]$, $\delta > 0$, then $\|x\|_\delta = \sup_{|s-t|\leq\delta} |x(s) - x(t)|$ is a pseudonorm on $C[0,1]$. One has the following criterion for tightness of a sequence of measures $\{\mu_n\}$ on $C[0, 1]$; see, e.g., **[1]**. The sequence $\{\mu_n\}$ is tight if and only if $\forall\, \varepsilon > 0$, $\exists\, A > 0$ and $\delta > 0$ such that

$$\sup_{n\geqq 1} \mu_n\{x: |x(0)| > A\} < \varepsilon \tag{1.2}$$

and

$$\sup_{n\geqq 1} \mu_n\{x: \|x\|_\delta > \varepsilon\} < \varepsilon. \tag{1.3}$$

2. The reproducing kernel Hilbert space (rkhs) of a measure in WM_0^2. Parts of the main result (Theorem 6) are essentially known. The rkhs of $\mu \in M_0^2$ is discussed by Kuelbs **[21]**. The simple implication that supp μ (topological support of μ) is contained in $\overline{H^B}$ does not seem to have been observed before. I am indebted to G. Kallianpur and G. Pisier for useful discussions concerning the proof of Theorem 6. G. Pisier made the observation **[31]** that the sequence $\{\xi_i\}$ in the WM_0^2 expansion given below may be chosen to be a martingale difference sequence.

THEOREM 6. *Let $\mu \in WM_0^2$. Then there exists a separable Hilbert space H (rkhs of μ) such that*

$$H \subset B. \tag{2.1}$$

$$\textit{For some } \alpha > 0, \qquad \|x\|_B \leqq \alpha \|x\|_H, \quad \forall\, x \in H. \tag{2.2}$$

$$\operatorname{supp} \mu \subset \overline{H^B} \qquad (= \textit{closure of } H \textit{ in } B \textit{ in the norm topology}). \tag{2.3}$$

Furthermore, if X is a B-valued r.v. in WM_0^2 then given any complete orthonormal system (CONS) $\{e_i\}$ in H there exists an orthonormal sequence $\{\xi_i\} \subset$

$L_2^0(\Omega, \mathscr{F}, P)$ (= the space of zero mean, square-integrable, real-valued r.v.'s. on $(\Omega, \mathscr{F}, P)$) such that

$$X \cong \sum_{i=1}^{\infty} e_i \xi_i, \tag{2.4}$$

where (2.4) *holds in the WM_0^2 sense, i.e.,*

$$E\left(\left|f(X) - \sum_{i=1}^{n} f(e_i)\xi_i\right|^2\right) \to 0 \quad \text{as } n \to \infty, \forall f \in B^*.$$

There exists a CONS $\{e_i\}$ in H such that the corresponding $\{\xi_i\}$ is an orthonormal martingale difference sequence, in which case if $X \in M_0^p$, $p \geqq 1$, then we have $E(\|X - \sum_{i=1}^n e_i \xi_i\|_B^p) \to 0$ as $n \to \infty$.

PROOF. Since $\mu \in WM_0^2$, it is clear that B^* is a subset of $L_0^2 = L_0^2(B, \mathscr{B}, \mu)$. Define U: $B^* \to L_0^2$ as $Uf = f$. If $f_n \to f$ in B^* norm, and $Uf_n \to g$ in L_2^0, i.e., $\int |f_n(x) - g(x)|^2 \, d\mu(x) \to 0$, then it is clear that $g = f (= Uf)$. Therefore U has closed graph and is a bounded linear operator from the Banach space B^* into L_0^2. Define

$$H_1 = \overline{U(B^*)}^{L_0^2}. \tag{2.5}$$

Let U^*: $L_0^2 \to B^{**}$ be the adjoint operator of U given by

$$\langle U^* \varphi, m\rangle = \langle \varphi, Um\rangle = \int \varphi m \, d\mu \tag{2.6}$$

for $\varphi \in L_0^2$, $m \in B^*$, and define

$$H = U^*(H_1) \subset B^{**}. \tag{2.7}$$

If $\varphi \in H_1$ and $U^*\varphi$ is the null functional on B^*, then (2.5) and (2.6) show that φ is the null element of H_1. Thus U^* is 1-1 on H_1 and if $f, g \in H$ with $U^*(f_1) = f$ and $U^*(g^1) = g$, then $(f, g)_H = \int f_1 g_1 \, d\mu$ is a well-defined inner product on H, and, by (2.7), H is a Hilbert space with this inner product. Since B is separable, L_2^0 is separable and so is H_1. Therefore H is separable.

We will now show that $H \subset B$. Let $h \in B^*$ and let $\{K_n\}$ be a nondecreasing sequence of convex compact subsets of B such that $\mu(K_n) \nearrow 1$ and define $h_n = hI_{k_n} - \beta_n$, where $\beta_n = \int_{k_n} h \, d\mu$, so that $h_n \in L_0^2$, $h_n \to h$ pointwise since $\beta_n \to 0$. By Proposition 1.1 **[29]** there is an element $\zeta_n \in B$ such that for $f \in B^*$

$$f(\zeta_n) = \int_{k_n} f(x)h(x) \, d\mu(x), \qquad n \geqq 1.$$

If $\{f_\alpha\}$ is a net in B^* such that $f_\alpha(x) \to 0$, $\forall\, x \in B$, then for n fixed, $\langle U^*h_n, f_\alpha\rangle = \int f_\alpha h_n \, d\mu(x) = f_\alpha(\zeta_n) \to 0$ in α. Hence $U^*h_n \in B$, $\forall\, n$. Since $h_n \to h$ in L_0^2, we have $U^*h_n \to U^*h$ in B^{**} which shows $U^*h \in B$. If $h \in H_1$, then there exist $g_k \to h$ in L_0^2, $g_k \in B^*$. By the previous argument $U^*g_k \in B$ and so does U^*h. This shows that $H \subset B$.

By (2.6) every member of B^* is in H^*; therefore (2.2) follows.

To see (2.3) suppose $f \in B^*$ and $f = 0$ on H; then $U^*f = 0$ and so $0 = \langle U^*f, f\rangle =$

$\int f^2\,d\mu$, which shows $f = 0$ on supp μ. It now easily follows from the Hahn-Banach theorem that supp $\mu \subset \overline{H^B}$.

Now let $X \in WM_0^2$ be a B-valued r.v. on $(\Omega, \mathscr{F}, P)$. Let $\mathscr{F}_1 = \sigma\{f(X), f \in B^*\}$, which is a sub-$\sigma$-algebra of $\mathscr{F}$. With $\mu = \mu_X$ let H and H_1 be defined as above. It is clear that the mapping $\psi(f) = f(X)$, $f \in B^*$, and extended by continuity to H_1, is an isometric isomorphism between H_1 and $L_0^2(\Omega, \mathscr{F}_1, P)$. If $f \in B^*$ we have the real orthogonal expansion of f in H_1 given by $f = \sum_{i=1}^{\infty} \tilde{e}_i \langle f, \tilde{e}_i \rangle_{H_1}$, where $\{\tilde{e}_i\}$ is a CONS in H_1. Let $\{e_i\}$ be the corresponding CONS in H and let $\psi(\tilde{e}_i) = \xi_i$, $i \geqq 1$. Then we have $f(X) = \sum_{i=1}^{\infty} \xi_i e_i$ in the sense of (2.4).

Finally, since $L_0^2(\Omega, \mathscr{F}_1, P)$ is a separable Hilbert space of mean 0, square-integrable real-valued random variables, we can pick a martingale difference sequence $\{\xi_i\}$ as a CONS in $L_0^2(\Omega, \mathscr{F}_1, P)$. Let $\{\tilde{e}_i\}$ and $\{e_i\}$ be the corresponding complete orthonormal systems in H_1 and H, respectively. The martingale convergence theorem for B-valued random variables **[2]** then gives the strong L^p convergence if $X \in M_0^p$. This completes the proof.

REMARKS. (1) If supp μ is known to be a linear space, then the argument for (2.3) works in the other direction as well and we have supp $\mu = \overline{H^B}$. In general, it will be hard to check this condition. On the other hand, if $\mu_m(A) = \mu(A+m)$ for $A \in \mathscr{B}$, $m \in H$, and the measures μ_m are all equivalent to μ (which happens in the Gaussian case), then again it is easy to show that supp $\mu \supset \overline{H^B}$ and one gets equality in (2.3). The above is thus an alternate proof of the known fact **[20]** that, for $\mu \in G$, supp $\mu = \overline{H^B}$.

(2) The part of the proof showing $H \subset B$ can be simplified by observing that it is enough to work with sequences rather than nets when B is separable. However, the above proof works without change when B is a locally convex linear topological space (possibly nonseparable) provided that μ has the property that $\mu(K_n) \uparrow 1$ for some nondecreasing sequence $\{K_n\}$ of convex compact sets. In this more general situation H need not be separable.

(3) If the random variables $\{\xi_i\}$ in (2.4) can be chosen to be independent, then a theorem of Itô and Nisio **[14]** shows that the series in (2.4) converges a.s. in norm. If in addition $X \in M_0^p$, $p \geqq 1$, then one has convergence in M_0^p.

3. Integrability of CLT r.v.'s. Theorem 3 and the fact that l^p, $p > 2$, is not of cotype 2 show that one can find examples in l^p, $p > 2$, where $X \in$ CLT and yet $E(\|X\|^2) = \infty$. Such an example was given in **[16]** with $B = C[0, 1]$. One may ask the question whether moments of some order exist if $X \in$ CLT. The following was shown in **[15]**.

THEOREM 7. *$X \in CLT \Rightarrow P[\|X\| > \lambda] = O(\lambda^{-2})$. In particular, $X \in CLT \Rightarrow E(\|X\|^{2-\varepsilon}) < \infty$ for every $\varepsilon > 0$.*

The same proof shows even more as noted by Pisier **[31]**. Recall that $S_n = X_1 + \cdots + X_n$, partial sum of the i.i.d. sequence $\{X_j\}$.

THEOREM 8. *If there exist $\varepsilon < \frac{1}{4}$ and $d > 0$ such that $\sup_{n\geqq 1} P[\|S_n/\sqrt{n}\| > d] \leqq \varepsilon$, then*

$$\sup_{n\geqq 1}\sup_{\lambda>0} \lambda^2 P[\|S_n/\sqrt{n}\| > \lambda] < \infty;$$

consequently $\sup_{n\geqq 1} E(\|S_n/\sqrt{n}\|^{2-\delta}) < \infty$ *for each* $\delta > 0$.

The proof is essentially a consequence of Lévy's inequality [**19**, p. 12]. The details can be found in either one of [**15**], [**31**].

4. Proofs of Theorems 2, 3 and 5. For the proof of Theorem 2 see [**13**]. A somewhat simplified version may be found in [**15**]. We will give details of the proofs of Theorems 3 and 5.

PROOF OF THEOREM 3. The implication ($\Leftarrow$) is proved in [**15**]. It is also a consequence of Proposition 3 (given below for the proof of Theorem 5). The proof of ($\Rightarrow$) is based on the following characterization of a Banach space of cotype 2: B is of cotype 2 if and only if given $\{x_j\} \subset B$, $\{\eta_j\}$ a standard Gaussian sequence (real-valued i.i.d. $N(0, 1)$ random variables), $\sum_{j=1}^\infty x_j\eta_j$ converges a.s. $\Rightarrow$ $\sum_{j=1}^\infty \|x_j\|^2 < \infty$. See [**27**] and [**31**] for a proof of this. It would thus suffice to prove the following

PROPOSITION 1. *Let* $\{x_j\} \subset B$ *and* $\{\eta_j\}$ *be a standard Gaussian sequence. Assume that* $Y = \sum_{j=1}^\infty \eta_j x_j$ *converges a.s. Then there exists a r.v.* $X \in CLT$ *such that* $\Gamma_X = \Gamma_Y$ *and* $E(\|X\|^2) = \sum_{j=1}^\infty \|x_j\|^2 \leqq +\infty$.

PROOF. On a suitable probability space $(\Omega, \mathscr{F}, P)$ define

$$X = \sum_{j=1}^{\infty} 2^{j/2}\, \eta_j x_j I(\Lambda_j) \tag{4.1}$$

where $\{\eta_j\}$ is a standard Guassian sequence, $\Omega = \bigcup_{j=1}^\infty \Lambda_j$, disjoint union, $P(\Lambda_j) = 2^{-j}$, and $\{\eta_j\}$ and $\{I(\Lambda_j)\}$ are two independent classes. It is clear that X is well defined and $E(\|X\|^2) = \sum_{j=1}^\infty \|x_j\|^2$, $\Gamma_X = \Gamma_Y$. We will now show that if $\{X_n\}$ is a sequence of independent copies of X, $S_n = X_1 + \cdots + X_n$, then $\{S_n/\sqrt{n}\}$ is a tight sequence.

For this purpose let $\{\eta_j^{(k)}\}$, $\{\Lambda_j^{(k)}\}$, $k \geqq 1$, be independent copies of $\{\eta_j\}$, $\{\Lambda_j\}$, respectively. We take $(\Omega, \mathscr{F}, P)$ as a product space, $\Omega = \Omega_1 \times \Omega_2$, $\mathscr{F} = \mathscr{F}_1 \times \mathscr{F}_2$, $P = P_1 \times P_2$ and regard $\{\eta_j^{(k)}\}$ on $(\Omega_1, \mathscr{F}_1, P_1)$, $\{\Lambda_j^{(k)}\}$ on $(\Omega_2, \mathscr{F}_2, P_2)$. Define for $\omega = (\omega_1, \omega_2)$

$$S_n(\omega_1, \omega_2) = \sum_{j=1}^{\infty} 2^{j/2}\, x_j \sum_{k=1}^{n} \eta_j^{(k)}(\omega_1) I(\Lambda_j^{(k)})(\omega_2).$$

Given $\varepsilon > 0$, pick $\varphi(n)$ such that

$$\sum_{j=\varphi(n)}^{\infty} n2^{-j} \leqq \varepsilon, \qquad \sum_{j=\varphi(n)-1}^{\infty} n2^{-j} > \varepsilon. \tag{4.2}$$

Let $\Delta_1 = \bigcup_{j=\varphi(n)}^\infty \bigcup_{k=1}^n \Lambda_j^{(k)}$. If $\omega_2 \notin \Delta_1$, then

$$S_n(\omega_1, \omega_2) = \sum_{j=1}^{\varphi(n)} 2^{j/2}\, x_j \sum_{k=1}^{n} \eta_j^{(k)}(\omega_1) I(\Lambda_j^{(k)})(\omega_2). \tag{4.3}$$

Now for any $K > 0$,

$$P_2\left[\bigcup_{j=1}^{\varphi(n)} \left\{\frac{1}{n}\sum_{k=1}^{n} I(\Lambda_j^{(k)}) > (K+1)2^{-j}\right\}\right] \leqq \sum_{j=1}^{\varphi(n)} \frac{2^j}{K^2 n} \tag{4.4}$$

by using Chebyshev's inequality with centered second moment. It is clear from the definition of $\varphi(n)$ in (4.2) that K can be chosen, depending only on ε and not on n, such that the probability in (4.4) is less than ε. Pick such a K and let Δ_2 be the corresponding set involved in (4.4). The point of all this is that if $\omega_2 \notin \Delta_1 \cup \Delta_2$ (a set whose measure is less than 2ε), then we may write

$$\frac{S_n(\cdot, \omega_2)}{\sqrt{n}} = \sum_{j=1}^{\varphi(n)} x_j \theta_j(\cdot, \omega_2),$$

where for fixed ω_2 the θ_j are independent Gaussian r.v.'s with $E_1(\theta_j^2) = 2^j \sum_{k=1}^{n} I(\Lambda_j^{k)})(\omega_2)/n \leqq K$ by (4.4). We now compare $S_n(\cdot, \omega_2)/\sqrt{n}$ with the Gaussian random variable $Y = \sum_{j=1}^{\infty} x_j \xi_j$ (defined on some probability space), where $\{\xi_j\}$ is a sequence of i.i.d. $N(0, K)$ random variables; the series for Y converges a.s. since $\sum_{j=1}^{\infty} x_j \eta_j$ converges a.s. by hypothesis. Regarding B as a closed subspace of $C[0, 1]$ and recalling that $\|x\|_\delta = \sup_{|s-t|\leqq\delta} |x(s) - x(t)|$ is a pseudonorm on $C[0, 1]$, we now apply **[18**, Theorem 5.3] to conclude that $E_1(\|S_n/\sqrt{n}\|_\delta) \leqq E(\|Y\|_\delta)$, provided that $\omega_2 \notin \Delta_1 \cup \Delta_2$. By the Fernique-Landau-Shepp result **[5]**, **[23]**, $E(\|Y\|) < \infty$; hence by dominated convergence $\lim_{\delta\to 0} E(\|Y\|_\delta) = 0$. It then follows via Chebyshev's inequality that $\{S_n(\cdot, \omega_2)/\sqrt{n}, \omega_2 \notin \Delta_1 \cup \Delta_2, n \geqq 1\}$ is a tight family of random variables. By Fubini's theorem the tightness of $\{S_n/\sqrt{n}\}$ follows.

REMARK. Pisier **[31]** remarks that D. Aldous has also proved a result similar to Theorem 3.

To prove Theorem 5 we first give the following result which is of some independent interest.

PROPOSITION 2. *$PG \subset M_0^1 \Rightarrow B$ is of cotype 2.*

PROOF. Let $\{x_j\} \subset B$, $\{\eta_j\}$ be a standard Gaussian sequence, and assume that $Y = \sum_{j=1}^{\infty} x_j \eta_j$ converges a.s. We then need to show that $\sum \|x_j\|^2 < \infty$ given that $\mathrm{PG} \subset M_0^1$. Let $\alpha_j > 0$, $\sum_{j=1}^{\infty} \alpha_j = 1$. Define $X = \pm x_j/\sqrt{\alpha_j}$ with probability $\alpha_j/2$. Note that $\Gamma_X(f, g) = \sum_{j=1}^{\infty} f(x_j)g(x_j) = \Gamma_Y$, $f, g \in B^*$; therefore $X \in \mathrm{PG}$. Since $\mathrm{PG} \subset M_0^1$, $E(\|X\|) = \sum_j \sqrt{\alpha_j}\, \|x_j\| < \infty$. This being true for all $\{\sqrt{\alpha_j}\} \in l^2$, we must have $\sum_{j=1}^{\infty} \|x_j\|^2 < \infty$.

PROOF OF THEOREM 5. If $\mathrm{PG} \subset \mathrm{CLT}$, then, by Theorem 7, $\mathrm{PG} \subset M_0^1$; hence, by Proposition 2, B must be of cotype 2.

Now assume B to be of cotype 2. To finish the proof we need a definition and a proposition.

DEFINITION. If B_1 and B_2 are Banach spaces, then a bounded linear operator $U: B_1 \to B_2$ is absolutely 2-summing if there is a constant c such that $\forall\, x_1, \cdots, x_n \in B_1$

$$\sum_{i=1}^{n} \|Ux_i\|_{B_2}^2 \leqq c \sup_{\|f\|\leqq 1; f\in B_1^*} \sum_{i=1}^{n} (f, x_i)^2.$$

PROPOSITION 3. *If B is of cotype 2 and $X \in PG$, then $\exists$ a Hilbert space H and a bounded linear operator $S: H \to B$ such that for some H-valued random variable Y with $E(\|Y\|_H^2) < \infty$, $X = SY$. (S, H depend only on the covariance of X.)*

PROOF. If $X \in \mathrm{PG}$, then $X \in WM_0^2$. By Theorem 6 we have the WM_0^2 expansion $X = \sum_j e_j \xi_j$, where $\{e_j\}$ is a complete orthonormal system in the rkhs H_1 of X and $\{\xi_i\}$ is an orthonormal sequence of real-valued random variables. Since $X \in$

PG, $\sum_j e_j \eta_j$ converges a.s. for a standard Gaussian sequence $\{\eta_i\}$ (Remark 3, §2). By a result of Maurey **[26]** and Linde and Pietsch **[24]** we then have a constant c_1 such that $\forall\, x_1, \cdots, x_n \in H_1$, $\eta_1, \cdots, \eta_n$ independent $N(0, 1)$,

$$(4.5) \qquad E\left(\left\|\sum_{i=1}^{n} x_i \eta_i\right\|_B^2\right) \leqq c_1 \sup_{\|x_i\|_H \leqq 1} \sum_{i=1}^{n} (f, x_i)^2.$$

In the terminology of **[24]** the property in (4.5) says that the continuous map $I: H_1 \to B$ is absolutely γ-summing. Since B is of cotype 2, we have another constant $c_2 > 0$ such that $\forall\, x_1, \cdots, x_n$ in B, $\eta_1, \cdots, \eta_n$ independent $N(0, 1)$,

$$(4.6) \qquad E\left(\left\|\sum x_i \eta_i\right\|_B^2\right) \geqq c_2 \sum_{i=1}^{n} \|x_i\|_B^2.$$

From (4.5) and (4.6) we see that the embedding $I: H_1 \to B$ is an absolutely 2-summing operator. By a fundamental factorization theorem of Pietsch **[30]** there is a Hilbert space H, a Hilbert-Schmidt mapping $U: H_1 \to H$ and a bounded linear mapping $S: H \to B$ such that $I = SU$. It is clear that if we set $Y = \sum_{i=1}^{\infty} U(e_i)\xi_i$, then Y is H-valued and $SY = X$; furthermore $E(\|Y\|_H^2) = \sum_{i=1}^{\infty} \|U(e_i)\|_H^2 < \infty$ since U is Hilbert-Schmidt.

The proof of Theorem 5 ($\Leftarrow$) can now be finished quickly. If $X \in$ PG, then $X = SY$ as in Proposition 3. Since $E(\|Y\|_H^2) < \infty$, Y satisfies the central limit theorem as an H-valued r.v.; since $S: H \to B$ is bounded linear, it is clear that $SY = X \in$ CLT (continuous maps take compacts into compacts).

We have the following obvious corollary of Proposition 3.

COROLLARY. *If B is of cotype 2 and $X \in PG$, then $X \in M_0^2$ and the WM_0^2 expansion for X obtained in Theorem 6 is an expansion in M_0^2.*

5. Conditions on X that imply $X \in$ CLT. The state of affairs here is rather unsatisfactory. A general result due to Pisier **[31]** is the following

THEOREM 9. $X_1 \in CLT \Leftrightarrow$ (i) $\sup_n E(\|S_n/\sqrt{n}\|) = \alpha(X_1) < \infty$, *and* (ii) $\forall\, \varepsilon > 0$, *there is a finite dimensional B-valued r.v. Y_1 such that $\alpha(X_1 - Y_1) < \varepsilon$.*

Note that condition (i) is equivalent to $\{S_n/\sqrt{n}\}$ being bounded in probability, i.e., $\forall\, \varepsilon > 0$, $\exists\, A > 0$ such that $\sup_n P[\|S_n/\sqrt{n}\| > A] < \varepsilon$. Condition (ii) in addition to (i) guarantees tightness. The proof of the implication ($\Leftarrow$) is essentially the same as that of Theorem 2. For ($\Rightarrow$) a martingale difference sequence expansion for X_1 (as given in Theorem 6) is useful **[31]**.

The following result in $C[0, 1]$ was given in **[17]**. Earlier work on this formulation is found in **[4]**, **[9]**.

THEOREM 10. *Let d be a metric on* $[0, 1]$ *continuous with respect to the usual metric. Suppose $\exists$ a real-valued r.v. M such that $E(M^2) < \infty$ and $\forall\, s, t \in [0, 1]$, $\omega \in \Omega$, $|X(s, \omega) - X(t, \omega)| \leqq M(\omega) d(s, t)$, where X is a $C[0, 1]$-valued r.v. Then*

$$\int_0 H_d(u)^{1/2}\, du < \infty \Rightarrow X \in CLT,$$

where $H_d(u) = \ln\{$minimal number of d-balls of radius $\leqq u$ that cover $[0, 1]\}$.

J. Zinn [35] has shown that this result can also be derived from Theorem 2. The details of his proof may also be found in [31].

By using a theorem of Delporte, M. Hahn [10] has shown that if $X_1(s)$ is a process with $E[X_1(s)] = 0$, $E[X_1^2(s)] < \infty$, $0 \leqq s \leqq 1$, and $\forall\, s, t \in [0, 1]$

$$(5.1) \qquad (E(X_1(s) - X_1(t))^2)^{1/2} \leqq f(|t - s|)$$

where $f(0) = 0$, $f \nearrow$, continuous on $[0, \varepsilon]$ for some $\varepsilon > 0$, then X_1 is actually a $C[0, 1]$-valued random variable provided that $\int_0 f(y)/y^{3/2}\, dy < \infty$. This condition in fact gives the same modulus of continuity for each member of the sequence $\{S_n/\sqrt{n}\}$, which insures tightness, hence the central limit theorem. She also shows by constructing counterexamples that the condition is very tight. The result has been further improved by Hahn and Klass [11]. G. Pisier has observed (personal communication) that if a covariance condition of the type (5.1) implies the continuity of X_1, then it also implies that $X_1 \in$ CLT. J. Devary [3] puts conditions on the distribution of X_1 to obtain results on the central limit theorem. A sample result may be found in [31].

6. Remarks. To understand what has been accomplished and what needs to be done, we give the following result of N. N. Vakhania [32], [33].

THEOREM 11. *Let $B = L_0^p(T, \Sigma, \mu)$, $1 \leqq p < \infty$, where (T, Σ, μ) is a σ-finite measure space. Then $X \in PG \Leftrightarrow X \in WM_0^2$ and $\int_T (E(X(t)^2))^{p/2}\, d\mu(t) < \infty$.*

PROOF. If $X \in$ PG, then $\exists\, \xi \in G$ such that $\Gamma_X = \Gamma_\xi$. By the Fernique-Landau-Shepp theorem

$$\infty > E(\|\xi\|_p^p) = E\int_T |\xi(t)|^p\, d\mu(t) = c_p \int_T (E(\xi(t)^2))^{p/2}\, dt$$
$$= c_p \int_T (E(X(t)^2))^{p/2}\, dt,$$

where the second equality comes from the fact that $\xi(t)$ is a centered Gaussian r.v. and the third one follows from $\Gamma_X = \Gamma_\xi$. Conversely assume that $X \in WM_0^2$ and satisfies the integral condition. We can construct a Gaussian process $\{\xi(t), t \in T\}$ on some probability space with $E(X(s)X(t)) = E(\xi(s)\xi(t))$, $\forall\, s, t \in T$. The above chain of equalities remains valid ($E(|\xi(t)|^p) = c_p(E(X(t)^2))^{p/2}$, hence $E(|\xi(t)|^p)$ is a measurable function of t) and it follows that ξ is indeed an $L_0^p(T, \Sigma, \mu)$-valued r.v., thus $X \in$ PG.

We get the following corollary of Theorems 5 and 11 since $L_0^p(T, \Sigma, \mu)$, $1 \leqq p \leqq 2$, is of cotype 2.

COROLLARY 1. *If $B = L_0^p(T, \Sigma, \mu)$, $1 \leqq p \leqq 2$, then $X \in CLT \Leftrightarrow \int_T (E(X(t)^2))^{p/2} dt < \infty$.*

We also get the following corollary of Zinn [35].

COROLLARY 2. *If μ is a probability measure on $B = C[0, 1]$, then μ regarded as a probability measure on $L^p[0, 1]$, $1 \leqq p < \infty$, belongs to CLT provided $\int x\, d\mu = \theta$, $\int \|x\|^2\, d\mu < \infty$.*

PROOF. For $1 \leqq p \leqq 2$, μ will clearly satisfy the integral condition of Corollary 1 and the result follows. For $p > 2$, $L^p[0, 1]$ is of type 2 and then the result follows from Theorem 2. The proof is terminated.

If $p > 2$, no necessary and sufficient conditions are known for $X \in$ CLT. More generally, as far as we know, no necessary and sufficient conditions are known for $X \in$ PG (or for $X \in$ CLT) for a space of type 2 or cotype 2. The question of characterizing domains of attraction of a Gaussian measure is one step further removed.

REFERENCES

1. P. Billingsley (1968), *Convergence of probability measures*, Wiley, New York and London. MR **38** #1718.
2. S. D. Chatterji (1964), *A note on the convergence of Banach-space valued martingales*, Math. Ann. **153**, 142–149. MR **28** #4583.
3. J. Devary (1975), *Regularity properties of second order processes*, Univ. of Minnesota, Ph.D. Thesis.
4. R. M. Dudley (1974), *Metric entropy and the central limit theorem in C(S)*, Ann. Inst. Fourier (Grenoble) **24**, 49–60.
5. X. Fernique (1970), *Intégrabilité des vecteurs gaussiens*, C. R. Acad. Sci. Paris. Sér. A-B **270**, A1698–A1699. MR **42** #1170.
6. R. M. Fortet and E. Mourier (1955), *Les fonctions aléatoires comme éléments aléatoires dans les espaces de Banach*, Studia Math. **15**, 62–79. MR **19**, 1202.
7. ——— (1965), *Resultats complementaires sur les éléments aléatoires prenant leurs valeurs dans un espace de Banach*, Bull. Sci. Math. **78**, 14–30.
8. D. J. H. Garling (1976), *Functional central limit theorems in Banach spaces*, Ann. Probability **4**, 600–611.
9. E. Giné (1974), *On the central limit theorem for sample continuous processes*, Ann. Probability **2**, 629–641. MR **51** #6921.
10. M. Hahn (1975), *Conditions for sample continuity and the central limit theorem* (preprint).
11. M. Hahn and M. J. Klass (1976), *Sample continuity of square integrable processes* (preprint).
12. J. Hoffmann-Jørgensen (1974), *Sums of independent Banach space valued random variables*, Studia Math. **52**, 159–186. MR **50** #8626.
13. J. Hoffmann-Jørgensen and G. Pisier (1976), *The strong law of large numbers and the central limit theorem in Banach spaces*, Ann. Probability **4**, 587–599.
14. K. Itô and M. Nisio (1968), *On the convergence of sums of independent Banach space valued random variables*, Osaka J. Math. **5**, 35–48. MR **38** #3897.
15. N. C. Jain (1975), *Central limit theorem in a Banach space*, Proc. First Conf. on Probability in Banach spaces (July 1975), Springer Lecture Notes in Math., Springer-Verlag, Berlin and New York, no. 526, pp. 113–130.
16. ——— (1976), *An example concerning* CLT *and* LIL *in Banach space*, Ann. Probability **4**, 690–694.
17. N. C. Jain and M. B. Marcus (1975), *Central limit theorem for C(S)-valued random variables*, J. Functional Analysis **19**, 216–231.
18. ——— (1975), *Integrability of infinite sums of independent vector-valued random variables*, Trans. Amer. Math. Soc. **212**, 1–36.
19. J. P. Kahane (1968), *Some random series of functions*, D. C. Heath, Boston, Mass. MR **40** #8095.
20. G. Kallianpur (1971), *Abstract Wiener processes and their reproducing kernel Hilbert spaces*, Z. Wahrscheinlichkeitstheorie und Verw. Gebiete **17**, 113–123. MR **43** #6961.
21. J. Kuelbs (1974), *An inequality for the distribution of a sum of certain Banach space valued random variables*, Studia Math. **52**, 69–87.
22. S. Kwapień (1972), *Isomorphic characterization of inner product spaces by orthogonal series with vector valued coefficients*, Studia Math. **44**, 583–595. MR **49** #5789.

23. H. J. Landau and L. A. Shepp (1970), *On the supremum of a Gaussian process*, Sankhyā Ser. A **32**, 369–378. MR **44** #3381.

24. W. Linde and A. Pietsch (1974), *Mappings of Gaussian cylindrical maesures in Banach spaces*, Theor. Probability Appl. **19**, 445–460.

25. E. Mourier (1953), *Eléments aléatoires dans un espace de Banach*, Ann. Inst. H. Poincaré **13**, 161–244. MR **16**, 268.

26. B. Maurey (1972), *Espaces de cotype p*, $0 < p \leqq 2$, Séminaire Maurey-Schwartz, 1972–1973.

27. B. Maurey and G. Pisier (1974), *Series de variables aléatoires vectorielles independentes et proprietés geometriques des espaces de Banach* (preprint).

28. K. R. Parthasarathy (1967), *Probability measures on metric spaces*, Academic Press, New York and London. MR **37** #2271.

29. R. R. Phelps (1966), *Lectures on Choquet's theorem*, Van Nostrand, Princeton, N. J. MR **33** #1690.

30. A. Pietsch (1966/67), *Absolut p-summierende Abbildungen in normierten Räumen*, Studia Math. **28**, 333–353. MR **35** #7162.

31. G. Pisier (1975), *Le theorème de la limite centrale et la loi du logarithme itere dans les espaces de Banach*, Séminaire Maurey-Schwartz, 1975–1976.

32. N. N. Vakhania (1965), *Sur une propriété des répartitions normales de probabilités dans les espaces* l_p $(1 \leqq p < \infty)$ *et H*, C. R. Acad. Sci. Paris **260**, 1334–1336. MR **30** #4282.

33. ——— (1976), *Probability distributions on linear spaces*, "Mecnereba", Tbilisi, 1971. (Russian)

34. S. R. S. Varadhan (1962), *Limit theorems for sums of independent random variables with values in a Hilbert space*, Sankhyā Ser. A **24**, 213–238. MR **30** #1536.

35. J. Zinn (1975), *A note on the central limit theorem in Banach spaces* (preprint).

UNIVERSITY OF MINNESOTA

Proceedings of Symposia in Pure Mathematics
Volume 31, 1977

A RENEWAL THEOREM FOR RANDOM WALK IN A RANDOM ENVIRONMENT*

HARRY KESTEN

ABSTRACT. A random environment is a sequence $\{\alpha_i\}_{-\infty<i<\infty}$ of i.i.d. random variables with values in [0, 1]. X_t denotes the position at each t of a particle which performs a random walk in this environment; $X_0 = 0$ and epoch unit of time the particle moves one unit to the left or to the right. When the environmental α_i are given and $X_t = i$, then X_{t+1} will be $X_t + 1$ ($X_t - 1$) with probability α_i (respectively $1 - \alpha_i$). We prove here that the environment as seen from the moving particle, i.e., the sequence $\{\alpha_{X_t+i}\}_{-\infty<i<\infty}$, has a limit distribution as $t \to \infty$.

1. Introduction. As in **[14]** we consider the following random walk in a random environment: Let $\{\alpha_i\}_{-\infty<i<\infty}$ be a doubly infinite sequence of independent random variables, each with the same distribution function, G say, which is concentrated on [0, 1]. These α_i represent the "random environment"; once this is chosen it remains fixed for all time. $\{X_t\}_{t=0,1,\cdots}$ is the random walk. It is a sequence of integer-valued random variables which change by $+1$ or -1 each unit of time. Let $\mathscr{A} = \sigma\{\alpha_i: -\infty < i < +\infty\}$ be the σ-field generated by the α_i. Then we take

$$(1.1)\qquad X_0 = 0, \qquad \begin{aligned} P\{X_{t+1} = X_t + 1 \mid \mathscr{A}, X_0,\cdots,X_t\} &= \alpha_i \text{ on } \{X_t = i\}, \\ P\{X_{t+1} = X_t - 1 \mid \mathscr{A}, X_0,\cdots,X_t\} &= \beta_i \equiv 1 - \alpha_i \text{ on } \{X_t = i\}. \end{aligned}$$

Thus, given the environment, X_t is a Markov chain. It can then be viewed as the sequence of states of a birth and death process with birth, respectively death, parameters α_i and $\beta_i = 1 - \alpha_i$. Note, however, that $\{X_t\}$ is not Markovian when $\{\alpha_i\}$ is not fixed. This model (in a slightly more complicated version) was introduced in **[5]** and **[21]** to represent certain crystallographical and biochemical processes.

AMS (MOS) subject classifications (1970). Primary 60J15, 60K05; Secondary 60J80.
Key words and phrases. Random walk in random environment, branching process in random environment, birth and death process, renewal theorem, regeneration points.
*Research supported by the NSF under grant MPS 72-04534A03.

Limit theorems for X_t were given in [**14**], [**15**] and [**19**]. Here we want to prove a limit theorem for the distribution of the environment as seen by an observer who moves with the random walk. I.e., if we think of X_t as the position of a particle at time t, then we consider an observer on the particle, who at time t is also at X_t. The value of α observed at his position at time t is α_{X_t} and the values of α immediately to his right and left are α_{X_t+1}, and α_{X_t-1}, etc. For such an observer, the environment at time t is described by the sequence

$$(1.2) \qquad A(t) = \{A_i(t)\}_{-\infty<i<\infty} \equiv \{\alpha_{X_t+i}\}_{-\infty<i<\infty}.$$

$A(t)$ is a (random) element of[1] $S \equiv [0, 1]^{\mathbf{Z}}$ and its distribution P_t is a probability measure on the product σ-field $\mathscr{S} = \prod_{i=-\infty}^{+\infty} \mathscr{B}_0$, where $\mathscr{B}_0$ is the Borel field of [0, 1]. We shall show that under suitable circumstances P_t converges weakly to a probability measure P_∞ on $\mathscr{S}$. The question of convergence of P_t was already raised in [**21**], but an additional motivation for studying it is given by the limit theorems of [**14**], [**15**] and [**19**]. It was shown in these references that if

$$(1.3) \qquad E \log \frac{1-\alpha_0}{\alpha_0} < 0 \quad \text{but } E\frac{1-\alpha_0}{\alpha_0} \geqq 1,$$

then

$$(1.4) \qquad X_t \to \infty \quad \text{but } \frac{1}{t} X_t \to 0 \text{ w.p.l.}$$

(See also Corollary 1 below.) Thus the particle in the random walk moves to $+\infty$, but more slowly than linearly. The intuitive explanation for this phenomenon is that there will be stretches in the environment where α_i is small, and hence also the rate at which the particle drifts to the right through such a stretch is small. Consequently the particle lingers in stretches with small α_i. As a result P_t will put more mass on small α's than P_0, the initial measure, which is nothing but $\prod_{i=-\infty}^{+\infty} G$. (Here, as well as in the sequel, we use the same symbol for a distribution function and the Borel measure on the line corresponding to it.) The existence of a limit P_∞ of P_t and a fairly explicit description of P_∞ in Corollary 3 and Theorem 2 make this precise.

There are two reasons to call the convergence of P_t a renewal theorem. One, a formal similarity with the ordinary renewal theorem when α_i can take only two values, say α' and α''. If we call the i's for which α_i equals α' "good", then the probability of being at a "good" point at time t is $P\{\alpha_{X_t} = \alpha'\}$, and our theorem implies that this probability has a limit. The second reason lies somewhat deeper and has to do with the use of regeneration points in the proof. In this paper we shall only deal with the case where $X_t \to \infty$ w.p.l. In this situation, the epochs t with the property

$$(1.5) \qquad X_m \geqq X_t > X_n \quad \text{for all } m \geqq t > n$$

are regeneration points, and limit theorems for processes with regeneration points are well known ([**10**, Chapter XI.8], [**17**], [**18**, §2]). These classical results immediately yield our Theorem 1 which deals with the "positive recurrent case" where the expected time between regeneration points is finite. Unfortunately, the case

[1] $\mathbf{Z}$ denotes the integers.

where this expectation is infinite, the so-called "null recurrent case" is considerably more complicated. This case is treated in Theorem 2, but we shall not give its very long proof here[2]. We now state our precise results after we introduce some notation.

$$\beta_i = 1 - \alpha_i, \qquad m_i = \beta_i/\alpha_i.$$

In Theorem 2 we assume the existence of a $\kappa \in (0, 1]$ such that

$$Em_0^\kappa = 1. \tag{1.6}$$

Such a κ is necessarily unique. We denote by G^κ the probability measure on $[0, 1]$ with $G^\kappa(dx) = ((1 - x)/x)^\kappa G(dx)$. We write $(S', \mathscr{S}')$ and $(S'', \mathscr{S}'')$ for the product measurable spaces $\prod_{i=-\infty}^{-1} ([0, 1], \mathscr{B}_0)$, respectively $\prod_{i=0}^{\infty}([0, 1], \mathscr{B}_0)$, and write $s' = (\cdots, s'_{-2}, s'_{-1})$, respectively $s'' = (s''_0, s''_1, s''_2, \cdots)$, for generic points of these spaces. We often identify $(S, \mathscr{S})$ with $(S', \mathscr{S}') \times (S'', \mathscr{S}'')$ in an obvious way, so that we may view (s', s'') as a point s of S with coordinates

$$\begin{aligned} s_i &= s_i'' \quad \text{if } i \geqq 0, \\ &= s_i' \quad \text{if } i < 0. \end{aligned} \tag{1.7}$$

μ' will be the product measure $\prod_{-\infty}^{-1} G$ on $\mathscr{S}'$ and if (1.6) holds, μ^κ is the product measure $\prod_{i=0}^{\infty} G^\kappa$ on $\mathscr{S}''$. Lastly, when we discuss weak convergence (to be denoted by $\Rightarrow$) of probability measures on $\mathscr{S}, \mathscr{S}', \mathscr{S}''$ we always assume that S, S', S'' are endowed with the product topology.

THEOREM 1. *If* $Em_0 < 1$ *and (consequently)* $-\infty \leqq E \log m_0 < 0$, *then there exists a probability measure* P_∞ *on* $\mathscr{S}$ *such that* $P_t \Rightarrow P_\infty$ *as* $t \to \infty$.

(Some properties of P_∞ are given in Corollary 3 at the end of §3.)

THEOREM 2. *Assume that*

$$-\infty \leqq E \log m_0 < 0 \tag{1.8}$$

and that there exists a $\kappa \in (0, 1]$ *for which* (1.6) *holds as well as*[3]

$$Em_0^\kappa \log^+ m_0 < \infty. \tag{1.9}$$

If the distribution of $\log m_0$ *(excluding its possible atom at* $-\infty$*) is nonarithmetic*[4], *then there exists a probability measure* P_∞ *on* $\mathscr{S}$ *such that* $P_t \Rightarrow P_\infty$ *as* $t \to \infty$. P_∞ *can be described as follows: For a certain probability measure* μ^* *on* $\mathscr{S}''$ *(to be defined in Lemma 3) which is absolutely continuous with respect to* μ^κ, *choose* $(s', s'') \in S' \times S''$ *according to the measure* $\mu' \times \mu^*$. *With* s *as in* (1.7) *there exists, for almost all* s, $[\mu' \times \mu^*]$, *a unique probability measure* $\{\pi_k\}_{-\infty<k<\infty} = \{\pi_k(s)\}_{-\infty<k<\infty}$ *on* $\boldsymbol{Z}$ *satisfying*[5]

$$\pi_k(s) = \pi_{k-1}(s)\, s_{k-1} + \pi_{k+1}(s)\,(1 - s_{k+1}), \qquad k \in \boldsymbol{Z}. \tag{1.10}$$

[2]Copies of the complete proof are available from the author upon request.

[3]For a real number a, a^+ denotes $\max(0, a)$.

[4]By this we mean that the group generated by $(-\infty, +\infty) \cap \operatorname{supp}(\log m_0)$ is dense in $(-\infty, +\infty)$.

[5] $\{\pi_k(s)\}$ is the stationary measure for the conditional random walk $\{X_t\}$ given $\mathscr{A}$ of (1.1) on the atom $\{\alpha_i = s_i, i \in \boldsymbol{Z}\}$ of $\mathscr{A}$.

Given s, choose the index L with the distribution $\pi_l(s)$, i.e., $P\{L = l|s\} = \pi_l(s)$. Then P_∞ is the distribution of the sequence $\{s_{i+L}\}_{-\infty<i<\infty}$.

ACKNOWLEDGEMENT. The author is indebted to Frank Spitzer for suggesting the investigation of a limit distribution for α_{X_t}.

2. Renewal theoretic preparations. *The assumption* $E \log m_0 < 0$ (which was made explicitly in Theorems 1 and 2) *is in force throughout this paper.* From **[15]** and **[19]** we know that *under this assumption* $X_t \to \infty$ *w.p.*1.

We also use the following *notational conventions throughout this paper*: N is an arbitrary nonnegative integer, K with a subscript is some finite strictly positive constant whose precise value is unimportant. D with or without subscript will also be used for constants, but they have to be chosen judiciously, usually so as to satisfy conditions involving other parameters. For any function f, $\|f\|$ denotes the supremum of $|f|$ over all permissible values of the arguments of f; $I[\ \ ]$ denotes the indicator function of the event between square brackets. Finally we want to stress that the index t always takes its values in the positive integers.

Our first task is to show that (1.5) really defines regeneration points. For this purpose we put $\tau_0 = 0$ and for $k \geqq 0$

$$\tau_{k+1} = \inf\{t > \tau_k: X_m \geqq X_t > X_n \text{ for all } m \geqq t > n\}. \tag{2.1}$$

We also introduce random variables ξ_k, the so-called "random tours", which describe the "piece of the path between τ_k and τ_{k+1}". Specifically[6]

$$\begin{aligned}\xi_k = (\tau_{k+1} - \tau_k, X(\tau_{k+1}) - X(\tau_k), \alpha(X(\tau_k) + i),\\ 0 \leqq i < X(\tau_{k+1}) - X(\tau_k), X_{t+1} - X_t, \tau_k \leqq t < \tau_{k+1}).\end{aligned} \tag{2.2}$$

We write l_k and d_k for the "length", respectively "displacement", of ξ_k, i.e.,

$$l_k = \tau_{k+1} - \tau_k, \qquad d_k = X(\tau_{k+1}) - X(\tau_k). \tag{2.3}$$

By virtue of (2.1), $l_k \geqq 1$, $d_k \geqq 1$, and the state space Ξ of the ξ_i is therefore the disjoint sum

$$\Xi \equiv \sum_{l,d \geqq 1} \{l\} \times \{d\} \times [0, 1]^d \times \{-1, +1\}^l.$$

LEMMA 1. *W.p.1 all the τ_k are finite. For each $k \geqq 0$, ξ_{k+1} is independent of $\{\alpha_i: i < 0\}$ and $\xi_0, \cdots, \xi_k, \xi_1, \xi_2, \cdots$ are identically distributed, each having the conditional distribution of ξ_0, given*

$$B_0 = \{X_t \geqq 0 \textit{ for all } t \geqq 0\}.$$

Also $\{\alpha_i: i < 0\}$ is independent of B_0. Finally, given B_0, $\{\alpha_i: i < 0\}$ and $\xi_0, \xi_1, \cdots$ are conditionally independent and the conditional distribution of each ξ_k, $k \geqq 0$, given B_0 is the same.

PROOF. In agreement with the above definition of B_0 put

$$B_x = \{X_t \geqq x \text{ for all } t \geqq T_x\}, \tag{2.4}$$

where

[6] For typographical reasons we shall occasionally write $X(t)$ for X_t. Similarly we use $\alpha(i)$ for α_i and later $T(x)$ for T_x, $U(i)$ for U_i, and $\tau(k)$ for τ_k.

(2.5) $$T_x = \inf\{t \geqq 0: X_t = x\}.$$

Then one can obviously express I_{B_x} as some function, ι say, of the sequence of random variables

$$\{X_{t+1} - X_t: t = T_{x+n}, T_{x+n} + 1, \cdots, T_{x+n+1}\}, \qquad n = 0, 1, \cdots.$$

Here the function ι does not depend on x. Moreover, a slight extension of the proof of Theorem (1.8) in **[19]** shows that the sequence of random variables

$$\{X_{t+1} - X_t: t = T_x, T_x + 1, \cdots, T_{x+1}\}, \qquad x = 0, 1, \cdots,$$

is stationary and ergodic. By [**4**, Proposition 6.31], the sequence I_{B^x}, $x = 0, 1, \cdots$, is therefore stationary and ergodic too, and by the ergodic theorem [**4**, Theorem 6.21]

(2.6) $$\lim_{n\to\infty} \frac{1}{n+1} \sum_{x=0}^{n} I_{B_x} = P\{B_0\} \quad \text{w.p.1.}$$

We must have $P\{B_0\} = P\{B_x\} = E\{P\{B_x \mid \mathscr{A}\}\} > 0$, for if not, then $P\{B_x \mid \mathscr{A}\} = 0$ w.p.1, and also for each s and x

$$P\{X_s = x \text{ and } X_{t+s} \geqq x \text{ for all } t \geqq 0 \mid \mathscr{A}\} = 0$$

for almost all choices of the environment. This would mean that X_t has to return to $(-\infty, x)$ every time after it hits x, which contradicts $X_t \to \infty$. Thus, indeed $P\{B_0\} > 0$, and B_x occurs for infinitely many x by (2.6). When B_x occurs with $x \geqq 0$, then $t = T_x$ satisfies (1.5) and equals some τ_l, whence all $\tau_k < \infty$.

The next claim of the lemma is intuitively clear, because $\{\alpha_i\}_{i<0}, \xi_0, \cdots, \xi_k$ depend only on α_i for $i < X_{\tau_{k+1}}$ and X_t for $t \leqq \tau_{k+1} = T(X(\tau_{k+1}))$[7] (see (2.1)), whereas ξ_{k+1} depends only on α_i for $i \geqq X_{\tau_{k+1}}$ and $X_t - X_{\tau_{k+1}}$ for $t > \tau_{k+1}$. However, some care is needed because the restriction $\{\tau_{k+1} = t\}$ involves all future $\{X_m - X_t\}_{m\geqq t}$, so that τ_{k+1} is not a stopping time w.r.t. any obvious σ-fields. However, the only restriction on the future implied by $\{\tau_{k+1} = t\}$ is that $X_m \geqq X_t$ for all $m \geqq t$, so that for any positive function h on Ξ we have a.e. on the set $\{\tau_{k+1} = t, X(\tau_{k+1}) = x\}$

$$\begin{aligned} &E\{h(\xi_{k+1}) \mid \{\alpha_i\}_{i<0}, \xi_0, \cdots, \xi_k\} \\ &\quad = E\{h(\xi_{k+1}) \mid \{\alpha_i\}_{i<x}, X_n, n \leqq t, \tau_{k+1} = t, X(\tau_{k+1}) = x, X_m \geqq x \text{ for } m \geqq t\} \\ (2.7) &\quad = E\{h(\xi_0) \mid \{\alpha_i\}_{i<0}, X_m \geqq 0 \text{ for } m \geqq 0\} \\ &\quad = E\{h(\xi_0) \mid X_m \geqq 0 \text{ for } m \geqq 0\} = E\{h(\xi_0) \mid B_0\}. \end{aligned}$$

This proves at the same time that ξ_{k+1} is independent of $\{\alpha_i\}_{i<0}, \xi_0, \cdots, \xi_k$, and has the distribution of ξ_0 given B_0, which is independent of k.

Similarly $\{\alpha_i\}_{i<0}$ is independent of B_0, because B_0 depends only on the steps $X_{t+1} - X_t$ taken when $X_t \geqq 0$.

The final statement about the joint conditional distribution of $\{\alpha_i: i < 0\}, \xi_0, \xi_1, \cdots$, given B_0, can also be proved by the above argument. Indeed

$$B_0 = \{X_t \geqq 0 \text{ for all } t \geqq 0\} = \{X_t \geqq 0 \text{ for } \tau_0 \leqq t < \tau_1\}$$

can be defined in terms of ξ_0 only, and is therefore independent of ξ_k, $k \geqq 1$. □

It will be necessary to describe certain aspects of the distribution of ξ_1 or condi-

[7]See footnote 6.

tional distribution of ξ_0 given B_0 by means of a branching process with immigration in a random environment, analogously to **[14]**. As in **[14]** we consider a process $Z_0, Z_1, \cdots$ with the following probability structure: First $\{\alpha_i\}_{i\geqq 0}$ are chosen independently and each with the distribution G. When the α_i are given, $\{Z_t\}_{t\geqq 0}$ forms an inhomogeneous branching process with one immigrant at each epoch. Specifically, $Z_0 = 0$, and when $\{\alpha_i\}_{i\geqq 0}, Z_0, \cdots, Z_t$ are given, then Z_{t+1} has the distribution of the sum of $(Z_t + 1)$ independent random variables $V_0, V_1, \cdots$, each of which has the geometric distribution

$$P\{V_j = r\} = \alpha_t \beta_t^r, \qquad r \geqq 0.$$

(Compare also [**1**, Chapter VI.5].) In accordance with the terminology of branching processes we call Z_t the number of particles of the tth generation, V_0 (V_i) the number of children of the immigrant at time t (respectively the ith particle in the tth generation, $1 \leqq i \leqq Z_t$). The offspring of the immigrant and of the Z_t particles of the tth generation make up the $(t + 1)$th generation. We copy from **[14]** the following definitions[8]:

$$\nu_0 = 0, \qquad \nu_{k+1} = \inf\{t > \nu_k: Z_t = 0\}, \tag{2.8}$$

$$W_k = \sum_{\nu_k \leqq t < \nu_{k+1}} Z_t \quad \Big(= \sum_{\nu_k < t < \nu_{k+1}} Z_t\Big), \tag{2.9}$$

$$\begin{aligned} U_i^x &= \#\{t < T_x: X_t = i, X_{t+1} = i - 1\} \\ &= \#\text{ of steps by } \{X_t\} \text{ from } i \text{ to } i-1 \text{ during } [0, T_x). \end{aligned} \tag{2.10}$$

Analogously, we put

$$\begin{aligned} U_i &= \#\ \{t: X_t = i, X_{t+1} = i - 1\} \\ &= \text{total number of steps by } \{X_t\} \text{ from } i \text{ to } i-1. \end{aligned}$$

LEMMA 2. *For any $N \geqq 0$ the conditional joint distribution of $\{l_{N-i}, d_{N-i}, 0 \leqq i \leqq N, \alpha(X(\tau_{N+1}) - j), 0 < j \leqq X(\tau_{N+1}), U^{X(\tau_{N+1})}_{X(\tau_{N+1})-r}, 0 \leqq r \leqq X(\tau_{N+1})\}$ given B_0, is the same as the distribution of $\{2W_i + (\nu_{i+1} - \nu_i), (\nu_{i+1} - \nu_i), 0 \leqq i \leqq N, \alpha_{j-1}, 0 < j \leqq \nu_{N+1}, Z_r, 0 \leqq r \leqq \nu_{N+1}\}$.*

PROOF. Let $x_0 = 0 < x_1 < \cdots < x_N$. Then the event $\{B_0 \cap X(\tau_i) = x_i, 0 < i \leqq N\}$ can be described in terms of the U's as follows:

$$\{U_{x_i} = 0, 0 \leqq i \leqq N, \text{ but } U_j \geqq 1 \text{ for all } 0 \leqq j \leqq x_N \text{ with } j \notin \{x_0, \cdots, x_N\}\}. \tag{2.11}$$

This is so, because x equals some X_{τ_i} if and only if X_t never moves to the left from x. If $U_{x_N} = 0$ and $x \leqq x_N$, then $U_x = U_x^{x_N}$, since $U_{x_N} = 0$ implies $X_t \geqq x_N$ for all $t \geqq T_{x_N}$. Thus (2.11) equals

$$\{U_{x_N} = 0, U_{x_i}^{x_N} = 0, 0 \leqq i < N, \text{ but } U_j^{x_N} \geqq 1 \text{ for all } 0 \leqq j \leqq x_N \text{ with } j \notin \{x_0, \cdots, x_N\}\}. \tag{2.12}$$

Moreover, on $B_0 \cap \{X_{\tau_1} = x_1\}$

$$\begin{aligned} \tau_1 = T_{x_1} &= \#\text{ of steps by } \{X_t\} \text{ to the left before } T_{x_1} \\ &\qquad + \#\text{ of steps by } \{X_t\} \text{ to the right before } T_{x_1} \\ &= 2\{\#\text{ of steps by } \{X_t\} \text{ to the left before } T_{x_1}\} + x_1 \\ &= 2 \sum_{0 \leqq j < x_1} U_j^{x_1} + x_1, \end{aligned}$$

[8] $\#\{\ \}$ denotes the number of elements of $\{\ \}$.

because when $\{X_t\}$ goes from 0 to x_1 it must take x_1 steps more to the right than to the left (cf. [**14**, §1]). Similarly, when $X(\tau_i) = x_i$, $X(\tau_{i+1}) = x_{i+1}$, $l_i = \tau_{i+1} - \tau_i = 2\sum_{x_i \leqq j < x_{i+1}} U_j^{x_{i+1}} + (x_{i+1} - x_i)$.

Now fix any integers $\bar{l}_i, \bar{d}_i \geqq 1$ and sets $\mathscr{C}_i \subset [0, 1]$ and put $x_0 = 0$, $x_i = \sum_{j=0}^{i-1} \bar{d}_j$. If the integers $u_i \geqq 0$ satisfy

$$u_{x_i} = 0,\ u_j \geqq 1 \quad \text{for } j \notin \{x_0, \cdots, x_N\} \text{ and}$$
$$\bar{l}_i = 2 \sum_{x_i \leq j < x_{i+1}} u_j + \bar{d}_i = 2 \sum_{x_i < j < x_{i+1}} u_j + \bar{d}_i, \tag{2.13}$$

then

$$P\{l_i = \bar{l}_i, d_i = \bar{d}_i, 0 \leqq i \leqq N, \alpha(X\tau_{N+1}) - j) \in \mathscr{C}_j, 0 < j \leqq X(\tau_{N+1}),$$
$$U_r^{X(\tau_{N+1})} = u_r, 0 \leqq r \leqq X(\tau_{N+1}) \mid B_0\}$$
$$= \frac{1}{P\{B_0\}} P\{\alpha(x_{N+1} - j) \in \mathscr{C}_j, 0 < j \leqq x_{N+1}, U_r^{x_{N+1}} = u_r, \tag{2.14}$$
$$0 \leqq r < x_{N+1}, U_{x_{N+1}} = 0\}.$$

But $\{U_{x_{N+1}} = 0\} = B_{x_{N+1}}$ is independent of $\alpha(x_{N+1} - j)$, $j > 0$, and $U_r^{x_{N+1}}$, $0 \leqq r < x_{N+1}$, because $B_{x_{N+1}}$ is defined in terms of steps $X_{t+1} - X_t$ with $t \geqq T_{x_{N+1}}$ and $X_t \geqq x_{N+1}$. In addition

$$P\{U_{x_{N+1}} = 0\} = P\{B_{x_{N+1}}\} = P\{B_0\} \quad \text{and} \quad U_{x_{N+1}}^{x_{N+1}} = 0 \qquad \text{(by (2.10))},$$

so that the right-hand side of (2.14) equals

$$P\{\alpha(x_{N+1} - j) \in \mathscr{C}_j, 0 < j \leqq x_{N+1}, U_r^{x_{N+1}} = u_r, 0 \leqq r \leqq x_{N+1}\}$$
$$= E\{P\{U_{(x_{N+1}-r)}^{x_{N+1}} = u(x_{N+1} - r), 0 \leqq r \leqq x_{N+1} \mid \mathscr{A}\}; \tag{2.15}$$
$$\alpha(x_{N+1} - j) \in \mathscr{C}_j, 0 < j \leqq x_{N+1}\}.$$

It was shown in [**14**, §1], that given $\alpha_s = \bar{\alpha}_s$, $0 \leqq s < x$, the conditional distribution of $U_x^x, U_{x-1}^x, \cdots, U_0^x$, is the same as that of $Z_0, Z_1, \cdots, Z_x$, given $\alpha_s = \bar{\alpha}_{x-1-s}$, $0 \leqq s < x$. This together with the fact that $\{\alpha(x_{N+1} - j), 0 < j \leqq x_{N+1}\}$ has the same distribution as $\{\alpha_{j-1}, 0 < j \leqq x_{N+1}\}$ implies that (2.15) equals

$$P\{\alpha_{j-1} \in \mathscr{C}_j, 0 < j \leqq x_{N+1}, Z_r = u(x_{N+1} - r), 0 \leqq r \leqq x_{N+1}\}$$
$$= P\{2W_{N-i} + (\nu_{N-i+1} - \nu_{N-i}) = \bar{l}_i, \nu_{N-i+1} - \nu_{N-i} = \bar{d}_i, 0 \leqq i \leqq N, \tag{2.16}$$
$$\alpha_{j-1} \in \mathscr{C}_j, 0 < j \leqq \nu_{N+1}, Z_r = u(x_{N+1} - r), 0 \leqq r \leqq \nu_{N+1}\}$$

(see (2.8), (2.9) and (2.13)). This proves the lemma, because both (2.14) and the right-hand side of (2.16) vanish when (2.13) fails. □

COROLLARY 1. *$P\{l_1 = 1\} > 0$. $El_1 < \infty$ if and only if $Em_0 < 1$. Finally,*

$$\lim_{t\to\infty} \sum_{k=1}^{\infty} P\{\tau_k = t\} = \lim_{t\to\infty} P\{\text{some } \tau_k = t\} = \frac{1}{El_1}. \tag{2.17}$$

PROOF. As we observed already, the distribution of ξ_1 is the same as the conditional distribution of ξ_0, given B_0. Thus by Lemma 2 with $N = 0$, l_1 has the distribution of $2W_0 + \nu_1$. It follows that

$$P\{l_1 = 1\} = P\{W_0 = 0, \nu_1 = 1\} = P\{\nu_1 = 1\}$$
$$= E\{P\{Z_1 = 0 \mid \mathscr{A}\}\} = E\alpha_0 > 0$$

(recall $E\log\alpha_0^{-1}(1-\alpha_0)<0$). Also, $E\nu_1<\infty$ by [**14**, Lemma 2][9], and (compare (2.7) and proof of (2.28) in [**14**])

$$\begin{aligned}EW_0 &= E\sum_{t=0}^{\nu_1-1} Z_t = E\{\#\text{ of particles born before } \nu_1\}\\ &= \sum_{t=0}^{\infty} E\{\text{total offspring of the immigrant at time } t;\ \nu_1>t\}\\ &= \sum_{t=0}^{\infty} P\{\nu_1>t\}\ E\{\text{total offspring of immigrant at time } 0\}\\ &= E\nu_1\sum_{k=0}^{\infty} E\prod_{i=0}^{k} m_i = E\nu_1\sum_{k=0}^{\infty}(Em_0)^{k+1}.\end{aligned}$$

The last sum and hence El_1 is finite if and only if $Em_0<1$. Lastly, (2.17) follows from the renewal theorem [**9**, Chapter XIII.10–11], since the occurrence of some τ_k at t is an aperiodic (delayed) recurrent event by Lemma 1 and $P\{l_1=1\}>0$. □

Note that by the strong law of large numbers

$$\lim_{k\to\infty}\frac{1}{k}\tau_k=\frac{1}{El_1}\quad \text{w.p.1.}$$

We shall talk about the *positive recurrent case* when this limit is strictly positive, i.e., when $Em_0<1$, and about the *null recurrent case* when this limit is zero, i.e., when $Em_0\geqq 1$. In the latter case we have

COROLLARY 2. *Under the hypotheses of Theorem* 2

$$\lim_{x\to\infty}x^{\kappa}P\{l_1\geqq x\}=\lim_{x\to\infty}x^{\kappa}P\{l_0\geqq x\mid B_0\}=C$$

for some constant $0<C<\infty$.

PROOF. As in Corollary 1

$$\begin{aligned}P\{l_1\geqq x\}&=P\{l_0\geqq x\mid B_0\}=P\{2W_0+\nu_1\geqq x\}\\&\geqq P\{W_0\geqq x/2\}\sim K_{11}(x/2)^{-\kappa},\ x\to\infty,\end{aligned}$$

for some $0<K_{11}<\infty$, by [**14**, Lemma 6]. On the other hand, $E\nu_1<\infty$ by [**14**, Lemma 2], so that for each $0<\varepsilon<1$

$$\begin{aligned}P\{2W_0+\nu_1\geqq x\}&\leqq P\{\nu_1\geqq\varepsilon x\}+P\{W_0\geqq(1-\varepsilon)x/2\}\\&=o(x^{-1})+(K_{11}+o(1))((1-\varepsilon)x/2)^{-\kappa},\qquad x\to\infty.\end{aligned}$$

The corollary now follows with $C=2^{\kappa}K_{11}$, because $\kappa\leqq 1$. □

The next lemma introduces the measure μ^* of Theorem 2. We shall omit the rather lengthy proof of this lemma because it is only needed for the null recurrent case. In a manner of speaking this lemma is an appendix to §3 of [**13**]. We remind the reader that $(S'',\mathcal{S}'')$ and μ^{κ} have been defined in the introduction.

LEMMA 3. *Assume that the hypotheses of Theorem* 2 *hold and define*

$$\eta_k=\sum_{t=k+1}^{\infty}\prod_{j=k}^{t-1}m_j,\tag{2.18}$$

$$\gamma(x)=\min\left\{n:\sum_{t=0}^{n}\prod_{j=0}^{t}m_j>x\right\}.\tag{2.19}$$

[9]The proof of Lemma 2 in [**14**] uses only $E\log m_0<0$.

Then for some constant $0 < K_0 < \infty$

$$P\{\gamma(x) < \infty\} = P\{\eta_0 > x\} \sim K_0 x^{-\kappa}, \qquad x \to \infty, \tag{2.20}$$

but for each fixed N

$$P\{\gamma(x) \leqq N\} = o(x^{-\kappa}), \qquad x \to \infty. \tag{2.21}$$

Moreover, there exists a probability measure μ^* *on* $\mathscr{S}''$ *which is absolutely continuous with respect to* μ^κ *and such that the conditional distribution of* $\{\alpha_{\gamma(x)-i}\}_{i\geqq 0}$, *given* $\gamma(x) < \infty$, *converges weakly to* μ^*, *as* $x \to \infty$. *I.e., for any bounded continuous function f on* $[0, 1]^{N+1}$ *we have*

$$\lim_{x\to\infty} E\{f(\{\alpha_{\gamma(x)-i}\}_{0\leqq i\leqq N}) \,|\, \gamma(x) < \infty\} = \int \mu^*(ds'')f(\{s_i''\}_{0\leqq i\leqq N}). \tag{2.22}$$

Finally, for any $\varepsilon > 0$ *there exists a* $D_0 = D_0(\varepsilon) < \infty$ *such that*

$$\limsup_{x\to\infty} x^\kappa\, P\Big\{\gamma(x) < \infty \text{ but } \sum_{0\leqq k\leqq \gamma(x)-D_0} \prod_{j=0}^{k} m_j > \varepsilon x\Big\} \leqq \varepsilon. \tag{2.23}$$

3. The positive recurrent case. This section is devoted to the *proof of Theorem* 1, which is almost standard renewal theory. Several sources contain closely related results (see for instance [**17**]). The notation of §2 is maintained, and the assumptions of Theorem 1 are in force. Let

$$\rho(t) = \max\{k : \tau_k \leqq t\}, \tag{3.1}$$

and let f be a bounded continuous function on S which depends on $\{s_i\}_{-N\leqq i\leqq N}$ only, for some finite N. We decompose $Ef(A(t))$ according to the values of $\rho(t)$, $\tau_{\rho(t)}$, $X(\tau_{\rho(t)})$ and $\xi_{\rho(t)-N}, \cdots, \xi_{\rho(t)-1}$. Note that by definition $\tau_{\rho(t)} \leqq t < \tau_{\rho(t)+1}$ and $X(\tau_{\rho(t)}) \leqq X_t < X(\tau_{\rho(t)+1})$, so that for $\rho(t) \geqq N$

$$X_t - N \geqq X(\tau_{\rho(t)}) - N \geqq X(\tau_{\rho(t)-N}).$$

Thus $\{A_i(t)\}_{-N\leqq i\leqq N}$ and $f(A(t))$ are determined by $X_t - X(\tau_{\rho(t)})$, $\xi_{\rho(t)-N}, \cdots, \xi_{\rho(t)-1}$ and α_j for $j \geqq X(\tau_{\rho(t)})$. Thus we may write

$$\begin{aligned} &Ef(A(t)) - E\{f(A(t));\, \tau_{N+1} > t\} \\ &\quad= \sum_{m=0}^{t}\sum_{n=m}^{t}\sum_{x,y\geqq 0}\sum_{k=1}^{\infty}\int\{\tau_k = m,\, \tau_{k+N} = n,\, \tau_{k+N+1} > t, \\ &\qquad\qquad X_n = x,\, \xi_k \in dz_1, \cdots, \xi_{k+N-1} \in dz_N, \\ &\qquad\qquad\qquad \{\alpha_{x+j}\}_{j\geqq 0} \in ds'',\, X_t - X_n = y\} \\ &\qquad\qquad\qquad \cdot f(\alpha_{x-N+y}, \cdots, \alpha_{x-1}, s_0'', \cdots, s_y'', \cdots, s_{y+N}''). \end{aligned} \tag{3.2}$$

Here we use the same symbol f whether we consider it as a function on S or as a function of $\{s_i\}_{-N\leqq i\leqq N}$ only; also $(\alpha_{x-N+y}, \cdots, \alpha_{x-1}, s_0'', \cdots, s_{y+N}'')$ is to be read as $(s_{y-N}'', \cdots, s_{y+N}'')$ in case $y \geqq N$. Now given $\tau_{k+N} = n \leqq t$ and $X_n = x$, the quantities $\{\alpha_{x+j}\}_{j\geqq 0}$, $l_{k+N} = \tau_{k+N+1} - \tau_{k+N}$ and $X_t - X_n$ are functions of $\xi_{k+N}, \xi_{k+N+1}, \cdots$ so that by Lemma 1 given $\tau_k = m, \tau_{k+N} = n, \xi_k = z_1, \cdots, \xi_{k+N-1} = z_N$, the conditional distribution of $\{\alpha_{x+j}\}_{j\geqq 0}$, l_{k+N} and $X_t - X_n$ is simply the conditional distribution of $\{\alpha_j\}_{j\geqq 0}$, l_0 and X_{t-n}, given B_0. Thus we may write the right-hand side of (3.2) as

$$\sum_{m=0}^{t}\sum_{n=m}^{t}\sum_{k=1}^{\infty}\sum_{x\geq 0}\int P\{\tau_k = m, \tau_{k+N} = n, X_n = x, \xi_k \in dz_1, \cdots, \xi_{k+N-1} \in dz_N\} \tag{3.3}$$
$$\cdot \sum_{y\geq 0} g(t-n, y, \alpha_{x-N+y}, \cdots, \alpha_{x-1}),$$

where

$$g(r, y, \bar{\alpha}_{x-N+y}, \cdots, \bar{\alpha}_{x-1}) \tag{3.4}$$
$$= \int P\{\{\alpha_j\}_{j\geq 0} \in ds'', l_0 > r, X_r = y \mid B_0\} f(\bar{\alpha}_{x-N+y}, \cdots, \bar{\alpha}_{x-1}, s_0'', \cdots, s_{y+N}'').$$

Another application of Lemma 1 shows that, given $\tau_k = m$, the conditional distribution of $\xi_k, \cdots, \xi_{k+N-1}, \tau_{k+N} - \tau_k = l_k + \cdots + l_{k+N-1}$ and $\{\alpha_j\}$ for $X(\tau_{k+N}) - N \leqq j < X(\tau_{k+N})$ is the same as the conditional distribution of $\xi_0, \cdots, \xi_{N-1}, l_0 + \cdots + l_{N-1}$ and $\{\alpha_j\}$ for $X(\tau_N) - N \leqq j < X(\tau_N)$, given B_0. This allows us to write (substitute r for $t - n$ and s for $n - m$)

$$Ef(A(t)) = E\{f(A(N)); \tau_{N+1} > t\} + \sum_{r=0}^{t}\sum_{s=0}^{t-r}\sum_{k\geq 1} P\{\tau_k = t - r - s\} \tag{3.5}$$
$$\cdot \sum_{y\geq 0} E\{g(r, y, \alpha_{X(\tau_N)-N+y}, \cdots, \alpha_{X(\tau_N)-1}); l_0 + \cdots + l_{N-1} = s \mid B_0\}.$$

By Corollary 1, the dominated convergence theorem, and the fact that

$$\begin{aligned}\sum_{r\geq 0}\sum_{y\geq 0} &|g(r, y, \alpha_{X(\tau_N)-N+y}, \cdots, \alpha_{X(\tau_N)-1})| \\ &\leqq \sum_{r\geq 0} \|f\|\, P\{l_0 > r \mid B_0\} = \|f\| E\{l_0 \mid B_0\} \\ &= \|f\| El_1 < \infty,\end{aligned}$$

the limit of (3.5) as $t \to \infty$ exists and equals

$$\frac{1}{El_1}\sum_{r\geq 0}\sum_{y\geq 0} E\{g(r, y, \alpha_{X(\tau_N)-N+y}, \cdots, \alpha_{X(\tau_N)-1}) \mid B_0\}. \tag{3.6}$$

This proves the existence of lim $Ef(A(t))$ for any function f which depends on finitely many coordinates only, and hence Theorem 1 follows. □

Using Lemma 2 we can describe P_∞ to some extent. Indeed (3.6) can by Lemma 2 be written as

$$\begin{aligned}&\frac{1}{El_1}\sum_{r\geq 0}\sum_{y\geq 0} E\{g(r, y, \alpha_{N-y-1}, \cdots, \alpha_0)\} \\ &= \frac{1}{El_1}\sum_{r\geq 0}\sum_{y\geq 0} E\{g(r, y, \alpha_{y-N}, \alpha_{y-N+1}, \cdots, \alpha_{-1})\} \\ &= \frac{1}{El_1}\sum_{r\geq 0}\sum_{y\geq 0} E\{f(\alpha_{y-N}, \cdots, \alpha_{y+N}); l_0 > r, X_r = y \mid B_0\}.\end{aligned} \tag{3.7}$$

Here we used the fact that given B_0, the variables $\{\alpha_j\}_{j<0}$ and the functions l_0 and X_r of $\xi_0, \xi_1, \cdots$ are conditionally independent, and that $\{\alpha_j\}_{j<0}$ is independent of B_0 (see Lemma 1). Thus we proved the following representation of P_∞:

COROLLARY 3. $\int f(s)P_\infty(ds) = (1/El_1)\sum_{r\geq 0} E\{f(A(r)); l_0 > r \mid B_0\}$.

REFERENCES

1. K. B. Athreya and P. E. Ney, *Branching processes*, Springer-Verlag, Berlin and New York, 1972. MR **51** #9242.

2. P. Billingsley, *Convergence of probability measures*, Wiley, New York and London, 1968. MR **38** #1718.

3. D. Blackwell, *Extension of a renewal theorem*, Pacific J. Math. **3** (1953), 315–320. MR **14**, 994.

4. L. Breiman, *Probability*, Addison-Wesley, Reading, Mass., 1968. MR **37** #4841.

5. A. A. Černov, *Replication of a multicomponent chain by the "lightning" mechanism*, Biofizika **12** (1967), 297–301 = Biophysics **12** (1967), 336–341.

6. K. L. Chung, *Markov chains with stationary transition probabilities*, 2nd ed., Springer-Verlag, New York, 1967. MR **36** #961.

7. N. Dunford and J. T. Schwartz, *Linear operators*, Vol. I, Interscience, New York, 1958. MR **22** #8302.

8. K. B. Erickson, *Strong renewal theorems with infinite mean*, Trans. Amer. Math. Soc. **151** (1970), 263–291. MR **42** #3873.

9. W. Feller, *An introduction to probability theory and its applications*, Vol. I, 3rd ed., Wiley, New York, 1968. MR **37** #3604.

10. ———, *An introduction to probability theory and its applications*, Vol. II, 2nd ed., Wiley, New York, 1971. MR **42** #5292.

11. A. M. Garsia and J. Lamperti, *A discrete renewal theorem with infinite mean*, Comment. Math. Helv. **37** (1962/63), 221–234. MR **26** #5630.

12. T. E. Harris, *First passage and recurrence distributions*, Trans. Amer. Math. Soc. **73** (1952), 471–486. MR **14**, 567.

13. H. Kesten, *Random difference equations and renewal theory for products of random matrices*, Acta Math. **131** (1973), 207–248.

14. H. Kesten, M. V. Kozlov and F. Spitzer, *A limit law for random walk in a random environment*, Composition Math. **30** (1975), 145–168.

15. M. V. Kozlov, *Random walk in a one-dimensional random medium*, Teor. Veroyatnost. i Primenen **18** (1973), 406–408 = Theory of Prob. and Appl. **18** (1973), 387–388. MR **47** #7818.

16. P.-A. Meyer, *Probability and potentials*, Blaisdell, Waltham, Mass., 1966. MR **34** #5119.

17. D. R. Miller, *Existence of limits in regenerative processes*, Ann. Math. Statist. **43** (1972), 1275–1282. MR **47** #1149.

18. W. L. Smith, *Renewal theory and its ramifications*, J. Roy. Statist. Soc. Ser. B **20** (1958), 243–302. MR **20** #5534.

19. F. Solomon, *Random walks in a random environment*, Ph. D. Thesis, Cornell Univ., 1972; Ann. Probability **3** (1975), 1–31. MR **50** #14943.

20. F. L. Spitzer, *Principles of random walk*, Van Nostrand Co., Princeton, N. J., 1964. MR **30** #1521.

21. D. E. Temkin, *One-dimensional random walks in a two-component chain*, Dokl. Akad. Nauk SSSR **206** (1972), 27–30 = Soviet Math. Dokl. **13** (1972), 1172–1176. MR **47** #2671.

CORNELL UNIVERSITY

Proceedings of Symposia in Pure Mathematics
Volume 31, 1977

ON PREDICTION PROCESSES

FRANK B. KNIGHT

The concept of a prediction process was developed by the author in [3], and subsequently reworked with important additions by P. A. Meyer [5] and by P. A. Meyer and M. Yor [6]. The object of the present paper is to give a unified approach to the subject which includes all of the known examples, but returns to the basically pragmatic emphasis of [3]. At the same time, we incorporate as much as possible of the "general theory of processes" from [4] and [5], since this seems to give the theory its most definitive form. One immediate consequence is that the topological and measure-theoretic aspects of the prediction process, which in [3] were treated simultaneously, are separated out. In purely measure-theoretic terms, the prediction process $Z(t)$ becomes two distinct processes $Z(t\pm)$, where $Z(t+)$ is optional and $Z(t-)$ is previsible. Only later in §2, by introduction of a suitable topology, does it follow that $Z(t-) = \lim_{s\to t-} Z(s+)$ and $Z(t+) = \lim_{s\to t+} Z(s-)$.

The reference here to "the prediction process" requires substantial justification. It was pointed out at the end of [5] that the same general conclusions apply to the space of all right-continuous paths with left limits in any polish space E, instead of the particular "Lipschitzian" space of [3]. Thus, in speaking of *the* prediction process, it is to be understood that all of the different possibilities will be shown to be equivalent in a fundamental sense. The medium for expressing this equivalence will be a particular measurable process $X(t)$, in such a way that a more accurate expression than "the prediction process" would be "the prediction process of $X(t)$."

The purpose of $X(t)$, however, is not simply to provide a means of exhibiting the equivalence of various prediction processes. On the contrary, we regard the purpose of the prediction processes to be as a means of studying $X(t)$. This is the viewpoint of [3], in contrast to that of [5] and [6] where the prediction process was an end in itself. In the present work (as also in [3]) all of the structure associated with the prediction process is regarded as purely auxiliary to the study of a given process $X(t)$. Without doubt, both facets of the subject have a legitimate role to play. But it

---

AMS (MOS) subject classifications (1970). Primary 60G05.

appears to us that the potential utility and significance of the prediction process depend upon how it relates to the study of other processes. Much of its potential utility derives from the fact that it is possible to impose a greater degree of general regularity in an auxiliary space than one would be justified in assuming a priori for a given process. The significance of the prediction process then hinges upon how this regularity can be brought to bear on the study of the given process.

1. A general definition of the prediction process of a measurable process. As in [3], let $X(t, w)$, $t \geqq 0$, be an $\mathscr{L} \times \mathscr{F}$-completion measurable process on $(\Omega, \mathscr{F}, P)$ with countably generated state space $(E, \mathscr{E})$, where $(\mathscr{L}, L)$ denotes the Lebesgue σ-field and measure on R^+, and the completion is with respect to $L \times P$. In this section we will enumerate the essential measure-theoretic components of a prediction process of $X(t)$ $(= X(t, w))$. No explicit topological assumptions or consequences will be adduced.

(i) We begin with an auxiliary measurable space $(\Omega^*, \mathscr{F}^*)$, and an adapted family $\mathscr{F}^*(t)$ of countably generated σ-subfields of $\mathscr{F}^*$ $(= \bigvee_{t \geqq 0} \mathscr{F}^*(t))$. It is assumed that $(\Omega^*, \mathscr{F}^*)$ is a tractable space, such as a compact metrizable space, or more generally a Borel subset of a polish space, in such a way that regular conditional probabilities exist over $(\Omega^*, \mathscr{F}^*)$.

(ii) Letting $\mathscr{F}(t_1, t_2)$ denote, as in [3], the σ-subfield of $\mathscr{F}$ generated by all integrals $\int_{t_1}^{s} f(X(\tau))\, d\tau$, $t_1 < s < t_2$, $f \in b(\mathscr{E})$, we assume the existence of a mapping $m^*\colon \Omega \to \Omega^*$ such that, for all $t > 0$, $(m^*)^{-1}(\mathscr{F}^*(t-)) = \mathscr{F}(0, t)$, where $\mathscr{F}^*(t-) = \bigvee_{s<t} \mathscr{F}^*(s)$ and the notation signifies the collection of inverses of sets in $\mathscr{F}^*(t-)$ under m^*. The mapping m^* provides the only direct connection between $X(t)$ and the prediction process.

(iii) Let $(L^*(t), \mathscr{L}^*(t))$, $0 < t \leqq \infty$, denote the space of all $\mathscr{L}$-measurable, $(E, \mathscr{E})$-valued paths $w(s)$, $0 \leqq s \leqq t$, with the σ-field $\mathscr{L}^*(t)$ generated by all definite integrals $\int_0^{\tau} f(w(s))\, ds$, $f \in b(\mathscr{E})$, $0 < \tau < t$, and set $(L^*, \mathscr{L}^*) = (L^*(\infty), \mathscr{L}^*(\infty))$. We assume the existence of a family θ_s^*, $s > 0$ ($\theta_0^* =$ the identity) of *generalized translation operators*: $\Omega^* \to \Omega^*$, $\mathscr{F}^*/\mathscr{F}^*$-measurable, such that for each $g \in b(\mathscr{L}^*(t))$ there is a $g^* \in b(\mathscr{F}^*(t-))$ and conversely for each $g^* \in b(\mathscr{F}^*(t-))$ there is a $g \in b(\mathscr{L}^*(t))$ for which

$$g(X(s + (\cdot))) = g^* \circ \theta_s^* \circ m^* \quad \text{for all } s \geqq 0,\ P\text{-a.s.} \tag{1.1}$$

REMARK. While this assumption is somewhat complicated to state, it is easily satisfied in practice, as will be noted below. In fact, the use of θ_t^* on Ω^* in place of the usual θ_t on Ω (which are often troublesome to define) constitutes a significant advantage of the prediction process over the given process $X(t)$.

(iv) Let $(H, \mathscr{H})$ denote the space of all probability measures h on $(\Omega^*, \mathscr{F}^*)$, with the σ-field generated by the family $\{E^h f^*;\ f^* \in b(\mathscr{F}^*)\}$, and let $P^*\colon P^*(S^*) = P((m^*)^{-1} S^*)$, $S^* \in \mathscr{F}^*$, be the probability on $\mathscr{F}^*$ induced by P on $\mathscr{F}(0, \infty)$. For $t \geqq 0$, let $\mathscr{F}^*(t+)$ denote $\bigcap_{\varepsilon > 0} \mathscr{F}^*(t + \varepsilon)$ with the inclusion of all subsets of P^*-null sets in $\mathscr{F}^*$. Then a *prediction process of* $X(t)$ is a pair $Z(t+)$, $t \geqq 0$, and $Z(t-)$, $t > 0$, of $(H, \mathscr{H})$-valued measurable processes on $(\Omega^*, \mathscr{F}^*)$ such that $Z(t+)$ is measurable over the $\mathscr{F}^*(t+)$-optional σ-field (as a function of (t, w^*)), $Z(t-)$ is measurable over the $\mathscr{F}^*(t+)$-previsible σ-field, and for any finite $\mathscr{F}^*(t+)$-optional stopping time T^* (respectively, any finite $\mathscr{F}^*(t+)$-previsible stopping time $T^* > 0$) we have

(1.2a) $$E(f^* \circ \theta^*_{T^*} \circ m^* | \mathscr{F}(0, T^*m^*+)) = (E^{Z(T^*+)}f^*) \circ m^*$$

or respectively,

(1.2b) $$E(f^* \circ \theta^*_{T^*} \circ m^* | \mathscr{F}(0, T^*m^*)) = (E^{Z(T^*-)}f^*) \circ m^*, \qquad f^* \in b(\mathscr{F}^*).$$

REMARKS. It will be shown immediately below that T^*m^* is an $\mathscr{F}(0, t+)$-stopping time (respectively, an $\mathscr{F}(0, t+)$-previsible stopping time) where $\mathscr{F}(0, t+)$ denotes $\bigcap_{\varepsilon>0} \mathscr{F}(0, t+\varepsilon)$ augmented by all subsets of P-null sets in $\mathscr{F}(0, \infty)$. Then $\mathscr{F}(0, T^*m^*+)$ is defined in the usual way relative to $\mathscr{F}(0, t+)$, while, in the second case, $\mathscr{F}(0, T^*m^*) = \bigvee_n \mathscr{F}(0, T_n+)$ where T_n is any increasing sequence of $\mathscr{F}(0, t+)$-stopping times with

$$P\left\{T_n < T^*m^*, \lim_{n\to\infty} T_n = T^*m^*\right\} = 1.$$

The existence of such a sequence T_n, and the independence of this definition from the particular sequence chosen, are well-known consequences of the definition of previsibility **[4]** (see also **[1]**).

The connection between $X(t)$ and $Z(t\pm)$ assumed in (1.2a) and (1.2b) would not be sufficient were it not for the fact that to every $\mathscr{F}(0, t+)$-stopping time T there corresponds a T^*, and conversely. We state and prove this as

LEMMA 1.1. (a) *For every $\mathscr{F}^*(t+)$-stopping time T^*, $T = T^*m^*$ is an $\mathscr{F}(0, t+)$-stopping time. If T^* is $\mathscr{F}^*(t+)$-previsible, then T is $\mathscr{F}(0, t+)$-previsible.*

(b) *Conversely, for each $\mathscr{F}(0, t+)$-stopping time T there exists an $\mathscr{F}^*(t+)$-stopping time T^* with $T = T^*m^*$. If T is previsible, then T^* is $\mathscr{F}^*(t+)$-previsible.*

PROOF. It is clear from (ii) that, for $t \geqq 0$, $(m^*)^{-1}(\mathscr{F}^*(t+)) = \mathscr{F}(0, t+)$. Thus if T^* is an $\mathscr{F}^*(t+)$-stopping time, T^*m^* is an $\mathscr{F}(0, t+)$-stopping time. Moreover, if T^*_n increases to T^*, P^*-a.s., then $T^*_n m^*$ increases to T^*m^*, P-a.s., and thus previsibility of T^* implies that of T^*m^*.

Conversely, let T be an $\mathscr{F}(0, t+)$-stopping time, and introduce the usual $T_n = (k+1)2^{-n}$ on $S_{k,n} = \{k2^{-n} \leqq T < (k+1)2^{-n}\}$, $k = 0, 1, \cdots$. Choosing sets $S^*_{k,n} \in \mathscr{F}^*((k+1)2^{-n})$ with $S_{k,n} = (m^*)^{-1}S^*_{k,n}$, we define inductively $T^*_n = (k+1)2^{-n}$ on $S^*_{k,n} - \bigcup_{j<k} S^*_{j,n}$, and $T^*_n = \infty$ on $\Omega^* - \bigcup_k S^*_{k,n}$. Then clearly T^*_n is an $\mathscr{F}^*(t+)$-stopping time, and $T_n = T^*_n m^*$. Setting $T^* = \limsup_{n\to\infty} T^*_n$, we see that $T = T^*m^*$ as required. Suppose, finally, that T is $\mathscr{F}(0, t+)$-previsible, and let T_n be a sequence of $\mathscr{F}(0, t+)$-stopping times with

$$P\left\{T_n < T \text{ or } T = 0, \lim_{n\to\infty} T_n = T\right\} = 1.$$

Writing $T_n = T^*_n m^*$, it is clear that

$$P^*\left\{T^*_n < T^* \text{ or } T^* = 0, \lim_{n\to\infty} T^*_n = T^*\right\} = 1,$$

which completes the proof.

Before turning to the equivalence problem for the prediction processes of $X(t)$, it is useful to review what is known about their existence. According to **[3]**, there always exists a prediction process with the following choice of objects (continuing the notation of **[3]**).

(i') $\Omega^* = \Omega' = \{(y_n(t)), n = 1, 2, \cdots (t \geqq 0): y_n(0) = 0, 0 \leqq y_n(t+s) - y_n(t) \leqq s \text{ for all } s > 0\}$.

$\mathscr{F}^* = \mathscr{F}'$ is the Borel σ-field of Ω' with the (compact metirzable) topology of uniform convergence for each n on compact time sets. $\mathscr{F}^*(t) = \mathscr{F}'(t)$ is generated by $\{y_n(s), s \leqq t\}$. (Alternatively, one can begin by mapping $(E, \mathscr{E})$ into [0,1] as in [5], which avoids the sequence $n = 1, 2, \cdots$, but is less natural for some purposes.)

(ii') $m^*(w) = (\int_0^t h_n(X(s, w)\, ds)$ where (h_n): $0 \leqq h_n \leqq 1$, $h_n \in b(\mathscr{E})$, is any fixed sequence generating $\mathscr{E}$ (in the case of $E \subset [0,1]$ we may simply use $h_n(x) = h_1(x) = x$).

(iii') $\theta_t^* = i_t : i_t(y_n(s)) = (y_n(t + s) = y_n(t))$.

(iv') It is shown that $Z(t+)$ $(= Z(t))$ and $Z(t-)$ exist and are unique up to a fixed P^*-null set. Moreover, in the (compact, metrizable) topology of vague (weak-*) convergence of measures on $(H, \mathscr{H})$, $Z(t)$ is right-continuous with left limits $Z(t-)$. Finally, although it will not be needed here, we recall that to each $h \in H$ there corresponds a $Z^h(t)$, in such a way that $Z(t) = Z^{P^*}(t)$, and the $Z^h(t)$ are all homogeneous strong-Markov processes with a single Borel transition function on $(H, \mathscr{H})$.

It is perhaps worthwhile to point out what is needed to generalize the method of [3]. The necessary features are

(a) $(\Omega^*, \mathscr{F}^*)$ should be "topological", so that regular conditional probabilities exist,

(b) $\theta_{s+t}^* = \theta_s^* \theta_t^*$, and θ_t^* is right-continuous in t for each $w^* \in \Omega^*$,

(c) there exists a sequence $0 \leqq f_n^* \in b(\mathscr{F}^*)$ such that

$$\left\{\lambda \int_0^\infty (\exp -\lambda t)\, f_n^* \circ \theta_t^*\, dt,\ 0 < \lambda \text{ rational}\right\}$$

has bounded pointwise linear closure equal to $b(\mathscr{F}^*)$ and such that convergence of the λ-supermartingales

$$E^{Z(t)} \int_0^\infty (\lambda \exp -\lambda s)\, f_n^* \circ \theta_s^*\, ds = E^*\left(\int_0^\infty (\lambda \exp -\lambda s) f_n^* \circ \theta_{s+t}^*\, ds \,\middle|\, \mathscr{F}^*(t+)\right)$$

along rationals $t \to t_0+$ for all t_0 implies the existence of unique $Z(t_0) \in H$ such that the limits are given by

$$E^{Z(t_0)} \int_0^\infty (\lambda \exp -\lambda s) f_n^* \circ \theta_s^*\, ds.$$

It is to be remarked that if $(\Omega^*, \mathscr{F}^*)$ is a Borel subset of a compact metric space (as in [**6**, p. 5]), and if (b) holds, then if (f_n^*) may be chosen continuous so that $\int_0^\infty (\lambda \exp -\lambda t)\, f_n^* \circ \theta_t\, dt$ are continuous in the compact topology and (f_n^*) has bounded pointwise linear closure $b(\mathscr{F}^*)$, then (c) will also hold. Indeed, by passing to a subsequence we may assume that $Z(t)$ converges vaguely to some $Z(t_0)$ and then letting $\lambda \to \infty$ it follows that

$$E^{Z(t_0)} f_n^* = E^*(f_n^* \circ \theta_{t_0}^* \,|\, \mathscr{F}^*(t_0 +)) \quad \text{for all } n,$$

where the right side is determined uniquely (for all t, a.s.) by the martingale limits. Introducing linear combinations, and passing to bounded pointwise limits, then shows that $Z(t_0)$ is uniquely determined and has the required properties.

The second construction of a prediction process is that of [**5**] and [**6**]. This requires less auxiliary structure, but more assumptions on $X(t)$. It is assumed that $X(t)$ is the canonical process on the space of all right-continuous paths with left limits in a polish space E, absorbed at a discrete point Δ. In this case, it is easy to see that $\mathscr{F}(0, t) = \mathscr{F}^0(t-)$ (the usual generated σ-field). Then there exists a prediction process with

(i″) $\Omega^* = \Omega$, $\mathscr{F}^*(t) = \mathscr{F}^0(t)$.
(ii″) $m^* = 1$ (the identity mapping).
(iii″) $\theta_t^* = \theta_t$ (the usual translation operators).

Under these conditions the existence of $Z(t\pm)$ is established, but it does not follow that $Z(t+)$ is right-continuous with left limits $Z(t-)$ when Ω is given the compact metric topology and $(H, \mathscr{H})$ the topology of vague convergence. Nevertheless, it is shown in [**6**, Lemma 3] that Ω may be retopologized as a Borel subset of a compact metric space in such a way that the Borel σ-fields are preserved and $Z(t+)$ becomes right-continuous with left limits $Z(t-)$ in the vague topology of $(H, \mathscr{H})$ obtained from the new topology on Ω. As in the case of [**3**], this topology is generated by a countable family $\{E^h f_n^*, f_n^* \in b(\mathscr{F}^*)\}$.

We turn now to the equivalence question. A satisfactory answer is provided almost immediately by the section theorems of P.-A. Meyer. We have

THEOREM 1.1. *Let* $(\Omega_i^*, \mathscr{F}_i^*, m_i^*, \theta_{i,t}^*, Z_i(t\pm))$, $i = 1$ *or* 2, *be two prediction processes of* $X(t)$. *For any* $g \in b(\mathscr{L}^*)$, *as in* (iii), *let* g_i^* *satisfy* (1.1) *on* Ω_i^* *for this* g. *Then*

$$P\{(E^{Z_1(t+)}g_1^*) \circ m_1^* = (E^{Z_2(t+)}g_2^*) \circ m_2^* \text{ for all } t \geqq 0\} = 1, \tag{1.3a}$$

and

$$P\{(E^{Z_1(t-)}g_1^*) \circ m_1^* = (E^{Z_2(t-)}g_2^*) \circ m_2^* \text{ for all } t > 0\} = 1. \tag{1.3b}$$

Moreover, for each $w_i^* \in \Omega_i^*$, *let* t *be called a discontinuity time of* $Z_i(t\pm, w_i^*)$ *if* $Z_i(t+, w_i^*) \neq Z_i(t-, w_i^*)$.

Then $P\{Z_1(t\pm) \circ m_1^*$ *and* $Z_2(t\pm) \circ m_2^*$ *have the same discontinuity times*$\} = 1$.

PROOF. The processes $Z_i(t+) \circ m_i^*$ are $\mathscr{F}(0, t+)$-optional, in view of Lemma 1.1 while $Z_i(t-) \circ m_i^*$ are $\mathscr{F}(0, t+)$-previsible. Consequently, if (1.3a) fails, the optional section theorem of Meyer [**4**, p. 149] shows the existence of an optional stopping time T with

$$P\{(E^{Z_1(T+)}g_1^*) \circ m_1^* \neq (E^{Z_2(T+)}g_2^*) \circ m_2^*\} > 0.$$

By Lemma 1.1 we can write $T = T_i^* m_i^*$ for $\mathscr{F}_i^*(t+)$-optional T_i^*, and then (1.2a) and (1.1) imply that both sides of the above inequality reduce to $E(g(X(T + (\cdot)))| \mathscr{F}(0, T+))$, which is a contradiction. Assertion (1.3b) is proved analogously, using the previsibility of $Z_i(t-) \circ m_i^*$ and the previsible section theorem [**4**, loc. sit.]. Finally, since $\mathscr{F}_i^*$ are countably generated, we may choose sequences $f_{i,n}^* \in b(\mathscr{F}_i^*)$ such that t is a discontinuity time of Z_i if and only if $E^{Z_i(t+)}f_{i,n}^* \neq E^{Z_i(t-)}f_{i,n}^*$ for some n. By assumption (iii) there exists, for each $f_{1,n}^*$, a $g_n \in b(\mathscr{L}^*)$, and then a $g_{2,n}^* \in b(\mathscr{F}_2^*)$, such that $f_{1,n}^*$ and $g_{2,n}^*$ correspond as in the first part of the theorem. Consequently, except on a fixed P-null set, if t is a discontinuity time of Z_1 at $w_1^* = m_1^*(w)$, $w \in \Omega$, then it is also a discontinuity time of Z_2 at $w_2^* = m_2^*(w)$. This shows that the discontinuity times of $Z_1(t\pm) \circ m_1^*$ are contained in those of $Z_2(t\pm) \circ m_2^*$, P-a.s. Reversing the roles of Z_1 and Z_2 completes the proof.

2. The topology of the prediction process. A curious deficiency of Theorem 1.1 is that it does not make obvious the fact that the times of discontinuity are a finite or countably infinite set, P-a.s. Of course, this follows immediately whenever there is a metric topology on $(H, \mathscr{H})$ such that $Z(t+) \circ m^*$ is right-continuous with left limits $Z(t-) \circ m^*$ for all $t > 0$, P-a.s. In fact, by the equivalence of times of discontinuity, it suffices that such a topology exist for a single choice of $Z(t\pm)$, and this is known from [3]. However, it also seems worthwhile to show that such a topology exists for *any* choice of $Z(t\pm)$, provided at least that one reduces the σ-fields $\mathscr{F}^*(t)$ appropriately. We have

THEOREM 2.1. *Let* $(\Omega^*, \mathscr{F}^*, m^*, \theta_t^*, Z(t\pm))$ *be any prediction process of* $X(t)$. *Then there exist subfields* $\mathscr{F}_1^*(t) \subset \mathscr{F}^*(t)$ *such that the restrictions* $Z_1(t\pm)$ *of* $Z(t\pm)$ *to measures on* $\mathscr{F}_1^* = \bigvee_t \mathscr{F}_1^*(t)$ *define a prediction process of* $X(t)$ *on the corresponding space* $(H_1, \mathscr{H}_1)$ *and on this space there is a separable metric topology for which*

$$P\{Z_1(t+) \circ m^* \text{ is right-continuous for } t \geqq 0, \\ \text{with left limits } Z_1(t-) \circ m^* \text{ for } t > 0\} = 1.$$

PROOF. For each $m > 0$, let $f_{m,n}(x_1, \cdots, x_m) \geqq 0$, $n = 1, 2, \cdots$, be a sequence of bounded uniformly continuous functions with bounded pointwise linear closure $b(R^m)$, and let (h_n) be a sequence in $b(\mathscr{E})$ with bounded, pointwise, linear closure equal to $b(\mathscr{E})$. We consider the countable family

$$h_j(t) = f_{n,m}\left(\int_0^{t_1} h_{j_1}(X(t+s))\, ds, \cdots \int_0^{t_m} h_{j_m}(X(t+s))\, ds\right)$$

for all $n, m > 0$, all $j_1, \cdots, j_m > 0$, and all *rational* $t_1, \cdots, t_m > 0$. It is easy to see that for each fixed $t_0 > 0$ the set

$$S(t_0) = \{h_j(0) \colon t_1, \cdots, t_n < t_0\}$$

has bounded, pointwise, linear closure equal to $\mathscr{F}(0, t_0)$. Accordingly, for each $h_j(0) \in S(t_0)$ let $h_j^* \in b(\mathscr{F}^*(t_0-))$ satisfy $h_j^* \circ \theta_t^* \circ m^* = h_j(t)$ for all $t \geqq 0$. The existence of such h_j^* is assured, of course, by (iii). Let $\mathscr{F}_1^*(t_0)$ be the σ-field generated by the bounded, pointwise, linear closure of these h_j^*, for each t_0. We assume that the h_j^* are chosen consistently as t_0 increases, hence are in correspondence with the total set of $h_j(0)$. It is clear that $(m^*)^{-1}(\mathscr{F}_1^*(t_0-)) = \mathscr{F}(0, t_0)$; hence (ii) is still satisfied. On the other hand, since (1.1) is preserved under linear combinations and bounded pointwise limits of the functions g, and since the functions $h_j(0) \in S(t_0)$, considered as functions of the paths $w(s)$ instead of $X(s)$, generate $\mathscr{L}^*(t_0)$ in the same sense, we see that (iii) also is preserved.

Finally, the processes

$$(E^{Z(t+)} h_j^*) \circ m^* = E(h_j(t) \,|\, \mathscr{F}(0, t+))$$

and

$$(E^{Z(t-)} h_j^*) \circ m^* = E(h_j(t) \,|\, \mathscr{F}(0, t)),$$

are respectively $\mathscr{F}(0, t+)$-optional and $\mathscr{F}(0, t+)$-previsible by (iv), and consequently they are the optional and previsible projections of $h_j(t)$, unique up to P-null sets, in the sense of [**4**, pp. 210 and 214]. But since $h_j(t)$ is continuous in t for every $w \in \Omega$, it follows immediately by use of the optional and previsible section theorems that the optional projection is right-continuous, and the previsible projection is left-continuous with right limits equal to the former. Consequently, in the uniform structure on $(H_1, \mathscr{H}_1)$ induced by the functions $E^z h_j^*$, the process $Z(t+)$ is right-continuous with left limits $Z(t-)$, P^*-a.s. But this structure is trivially metrizable, and since the h_j^* generate $\mathscr{F}_1^*$, the $E^z h_j^*$ generate $\mathscr{H}_1$. This completes the proof.

REFERENCES

1. K. L. Chung and J. L. Doob, *Fields, optionality, and measurability*, Amer. J. Math. **87** (1965), 397–424. MR **35** #4972.

2. C. Dellacherie and P. -A. Meyer, *Probabilités et potentiel.* Chaps. I à IV, rev. ed., Publ. Inst. Math. Univ. Strasbourg XV, Hermann, Paris, 1975.

3. F. Knight, *A predictive view of continuous time processes*, Ann. Probability **3** (1975), 573–596.

4. P. -A. Meyer, *Guide détaillé de la théorie "générale" des processus*, Séminaire de Probabilités. II (Univ. Strasbourg, 1967), Lecture Notes in Math., vol. 51, Springer-Verlag, Berlin and New York, 1968, pp. 140–165. MR **38** #3911.

5. ———, *La théorie de la prediction de F. Knight*, Séminaire de Probabilités. X (Univ. Strasbourg), Lecture Notes in Math., Springer-Verlag, Berlin and New York, 1976, pp. 86–103.

6. P. -A. Meyer and M. Yor, *Sur la théorie de la prediction, et le probleme de decomposition des tribus* $\mathscr{F}_{t+}^0$, Séminaire de Probabilités. X (Univ. Strasbourg) Lecture Notes in Math., Springer-Verlag, Berlin and New York, 1976, pp. 104–117.

UNIVERSITY OF ILLINOIS

Proceedings of Symposia in Pure Mathematics
Volume 31, 1977

A DERIVATION OF THE BOLTZMANN EQUATION FROM CLASSICAL MECHANICS*

OSCAR E. LANFORD III

Consider a system of n elastic spheres of diameter d, contained in a region Λ of Euclidean space. The spheres move freely except for elastic collisions with each other and with the boundary of Λ. Take an initial phase point in which the particles have the centers located at $q_1, \cdots, q_n \in \Lambda$ and have velocities $v_1, \cdots, v_n$, and describe the phase point by its one-particle distribution $n^{-1} \sum_{i=1}^{n} \delta_{q_i, v_i}$ which is a discrete probability measure on the one-particle phase space $\Lambda \times \boldsymbol{R}^3$. Suppose that this discrete measure is near, in the sense of the weak topology for measures, to some absolutely continuous probability measure $f(q, v)\, dq dv$. What can we then say about the one-particle distribution at later times?

It is believed that, at least at small densities, a continuous approximation to the discrete one-particle distribution at later times is given by the solution to the Boltzmann equation, a nonlinear evolution equation for probability densities which may be written as:

$$\frac{d}{dt} f_t = -v \cdot \operatorname{grad}_q f_t$$

$$+ nd^2 \int dv_1 \int d\hat{w}\, \hat{w} \cdot (v - v_1)\{f_t(q, v') f_t(q, v_1') - f_t(q, v) f_t(q, v_1)\}$$

where (as above) n is the number of particles, d is their diameter, where the $\hat{w}$ integration is over the hemisphere $\{\hat{w}: |\hat{w}| = 1,\ \hat{w} \cdot (v - v_1) \geqq 0\}$, and where v', v_1' are the outgoing velocities after an elastic collision with incoming velocities v, v_1 and momentum transfer in the direction $\hat{w}$. We will formulate here a theorem which says that the Boltzmann equation does indeed give a description of the motion of a typical phase point of a dilute gas of elastic spheres which becomes exact in the limit as the size of the particles goes to zero and their number goes to infinity, at

AMS (MOS) subject classifications (1970). Primary 82A40.

*Research was supported in part by NSF grant GP-42225.

least for a period of time of the order of one-fifth of the mean time between collisions. The exposition will be concise; a detailed discussion of the motivation may be found in [2] and an alternative formulation of the theorem together with an outline of the proof in [3]. An extension to more general finite-range repulsive interactions has been given in [1].

The limit $n \to \infty$, $d \to 0$, will be taken is such a way that the quantity nd^2 appearing in the Boltzmann equation approaches a finite nonzero limit; this corresponds schematically to holding the mean free path fixed. For convenience only, we keep the container Λ fixed and of unit volume as we pass to the limit. For given n, d, we let $\mathfrak{X}(n, d)$ denote the phase space for the n particle system, i.e.,

$$\mathfrak{X}(n, d) = \{(q_1, v_1; \cdots; q_n, v_n) \in (\Lambda \times \boldsymbol{R}^3)^n : |q_i - q_j| \geqq d \text{ for } i \neq j\}.$$

For each subset Λ of the one-particle phase space $\Lambda \times \boldsymbol{R}^3$, we define a function F_Δ on $\mathfrak{X}(n, d)$ which gives the fraction of the particles in Δ, i.e.,

$$F_\Delta(x_1, \cdots, x_n) = \frac{1}{n} \sum_{i=1}^{n} \varphi_\Delta(x_i)$$

where x_i denotes (q_i, v_i) and φ_Δ the characteristic function of Δ.

Now let f denote a probability density on $\Lambda \times \boldsymbol{R}^3$. A sequence (n_j, d_j, μ_j) where

(i) (n_j) is a sequence of positive integers approaching infinity,

(ii) (d_j) is a sequence of positive real numbers approaching zero,

(iii) for each j, μ_j is a probability measure on $\mathfrak{X}(n_j, d_j)$ which is symmetric under interchange of particles,

will be said to be an *approximating sequence* for f if for each rectangle Δ in $\Lambda \times \boldsymbol{R}^3$ the random variable F_Δ on $(\mathfrak{X}(n_j, d_j), \mu_j)$ converges in distribution to the constant $\int_\Delta f(x)\, dx$ as $j \to \infty$. Intuitively, this means that μ_j assigns high probability to finding the (random) one-particle distribution $n_j^{-1} \sum_{i=1}^{n_j} \delta_{x_i}$ near the continuous measure $f(x)\, dx$.

Consider now an approximating sequence (n_j, d_j, μ_j) for f such that $n_j d_j^2$ converges to a nonzero limit. We say that the approximating sequence has *the Boltzmann property* (*to time* τ) if for each $t \in [0, \tau)$ the sequence $(n_j, d_j, \mu_j(t))$ is an approximating sequence for a probability density f_t and if f_t is a solution to the Boltzmann equation (with nd^2 replaced by the limit of $n_j d_j^2$). With this terminology, a derivation of the Boltzmann equation takes the form of a theorem asserting that members of some class of approximating sequences have the Boltzmann property.

To state such a theorem, we need a little more terminology. If μ is a symmetric probability measure on $\mathfrak{X}(n, d)$ which is absolutely continuous with respect to Lebesgue measure and which therefore has the form $\mu(x_1, \cdots, x_n)\, dx_1 \cdots dx_n$, we define its kth *rescaled correlation function* by

$$f_k(x_1, \cdots, x_k) = \frac{n!}{(n-k)!} \frac{1}{n^k} \int dx_{k+1} \cdots dx_n\, \mu(x_1, \cdots, x_n).$$

We will write $f_k^{(j)}$ for the kth rescaled correlation function of μ_j. It is easy to verify that (n_j, d_j, μ_j) is an approximating sequence for f if and only if

$$\lim_{j \to \infty} f_k^{(j)}(x_1, \cdots, x_k) = f(x_1) \cdots f(x_k), \qquad k = 1, 2, \cdots,$$

in the sense of weak-* convergence of measures.

We now have

THEOREM. *Let* (n_j, d_j, μ_j) *be an approximating sequence for* f. *Assume*:

(1) *There exist constants* $z > 0$, $\beta > 0$, C *such that*

$$f_k^{(j)}(x_1, \cdots, x_k) \leqq C \prod_{i=1}^{k} \rho_{z,\beta}(v_i)$$

for all $j, k, x_1, \cdots, x_k$, *where*

$$\rho_{z,\beta}(v) = z\left(\frac{m\beta}{2\pi}\right)^{3/2} \exp\left[-\frac{m\beta}{2} v^2\right].$$

(2) f *is continuous, and for each* $k = 1, 2, \cdots$ *the convergence of* $f_k^{(j)}(x_1, \cdots, x_k)$ *to* $\prod_{i=1}^{k} f(x_i)$ *is uniform on compact sets in* $\{(q_1, v_1; \cdots; q_n, v_n): q_l \neq q_m \text{ for } l \neq m\}$.

Then (n_j, d_j, μ_j) *has the Boltzmann property to time*

$$\tau = .2\left(\frac{m\beta}{3}\right)^{1/2} \frac{1}{\pi(nz)d^2}.$$

The time τ may be interpreted as follows. Condition (1) says that the rescaled correlation functions are majorized by the rescaled correlation functions for an ideal gas with density $n_j z$ and temperature $(k\beta)^{-1}$, where k is Boltzmann's constant. The mean free time for a low density gas in equilibrium is given approximately by $(m\beta/3)^{1/2}\ (1/\pi(n_j z)d_j^2)$.

Thus the time τ should be roughly equal to one-fifth of the mean free time (or perhaps of some minimum local mean free time) for the system under consideration. It should perhaps be noted that there seems to be no reason not to believe that an approximating sequence of the sort considered has the Boltzmann property for all positive times, but presently available techniques of proof work only for the comparatively short times indicated.

REFERENCES

1. F. G. King, *The* BBGKY *hierarchy for positive potentials*, Ph. D. Dissertation, Dept. of Math., Univ. of Calif., Berkeley, 1975 (to appear).

2. O. E. Lanford, *On a derivation of the Boltzmann equation*, Proc. Internat. Conf. on Dynamical Systems in Mathematical Physics, Rennes, 1975, Astérisque (to appear).

3. ———, *Time evolution of large classical systems*, Dynamical Systems, Theory and Applications (J. Moser, editor), Lecture Notes in Physics, vol. 38, Springer-Verlag, Berlin and New York, 1975, pp. 1–111.

UNIVERSITY OF CALIFORNIA, BERKELEY

Proceedings of Symposia in Pure Mathematics
Volume 31, 1977

RANDOM TIMES AND DECOMPOSITION THEOREMS

P. WARWICK MILLAR*

1. Introduction. Let $\{X_t, t > 0\}$ be a right continuous Markov process. Much effort in general Markov theory has dwelt on the problem of isolating those random times R such that (a) the post R process $\{X(R + t), t > 0\}$ has a Markovian structure and/or (b) the post R process is conditionally independent of events "before R", given X_R. The most celebrated decomposition of the form (a), (b) is the strong Markov property, according to which if R is a stopping time, the post R process is Markov with the same transitions as the original process. Conditions under which the strong Markov property holds were investigated in the 1950's by Doob **[8]**, Hunt **[14]**, Blumenthal **[1]**, Dynkin **[11]** and others. In the 1960's it was noticed (**[3]**, **[4]** for example) that analogous decompositions could be proved for a class of random times of which the paradigm case is the "time of last exit from a fixed set". In this case the post R process is Markov, but its transitions are different from those of the original process. Despite the added complication, the decomposition still yields a good deal of insight since the transition functions of the post R process have a transparent probabilistic interpretation. Decomposition at "last exit" and related times has received nearly definitive formulation in the papers of Meyer-Smythe-Walsh **[40]**, Pittinger-Shih **[33]**, Getoor-Sharpe **[12]**, **[13]** and Maisonneuve-Meyer **[23]**.

In the late 1960's, D. Williams **[41]** noticed that a decomposition theorem at R could be proved for certain diffusions, if R was the time at which the process reached its ultimate minimum; this decomposition was a key but difficult step in his deep investigation of Brownian local time **[42]**, **[24]**. Further decompositions at minima and closely related time have been established by Jacobsen **[17]** (diffusions), Pitman **[34]** (conditioned Brownian motion) and Millar **[29]** (general Markov pro-

AMS (MOS) subject classifications (1970). Primary 60J25, 60J40; Secondary 60F99, 60J60, 60J30.

*Research supported by NSF grant MPS75 10376.

cesses). In this paper a broad class of random times (called randomized coterminal times) is propounded, which includes last exists and minima as special cases. In concrete examples these times amount to "the last time that the process leaves a set picked at random". The decomposition at minima, and at these more general times, has the feature that the post R process is not Markov in its own right, but, conditional on appropriate random variables, it is Markov with transitions that depend on the values of the conditioning variables. Again, despite this seeming complication, much insight is gained from the decomposition because of the probabilistic interpretation of these post R transition functions.

This paper surveys what is currently known about decomposition theorems, with most of the attention focussed on the problems associated with randomized coterminal times. The decomposition in §3 is new; proof will appear elsewhere. §2 presents some basic examples of random times; in §3 is the formal definition of a randomized coterminal time R together with the Markovian character of the post R process. §4 discusses analogues of the Blumenthal zero-one law for the post R process. In §5 conditional independence of pre and post R processes is proved for various R; some arguments use the zero-one laws of §4. Finally §6 illustrates the results of previous sections by decomposing the path of a Levy process into a number of interesting fragments.

2. Random times. This section contains basic notions involving general random times, together with a number of examples. The examples after (2.11) are intended to give some feel for the formal definition of randomized coterminal time given in §3. We begin with a description of the basic framework.

Let $\{X_t, \mathscr{F}_t, t > 0\}$ be a right continuous, left limit strong Markov process with state space $(E, \mathscr{E})$, a locally compact separable metric space. The sigma fields $\mathscr{F}_t$ are assumed to be right continuous completions of the natural sigma fields $\mathscr{F}_t^0 = \sigma\{X_s, s \leqq t\}$, as described in the usual general theory of Markov processes **[2]**, **[26]**; here and throughout the paper $\sigma\{\cdots\}$ denotes the sigma field generated by whatever appears between the braces. Set $\mathscr{F} = \bigvee \mathscr{F}_t$. A point Δ has been set aside as the usual terminal absorbing state: $X_\infty = \Delta$ and if $\zeta = \inf\{t: X_t = \Delta\}$, then $X_t = \Delta$ for all $t > \zeta$. The transition operators P_t are assumed to form a Hunt semigroup; there are then available the usual measures P^x, $x \in E$, on the space Ω of paths of X so that under P^x, $Y_t(\omega) = \omega(t)$ is Markov with transitions P_t and starting point x. We shall often take the canonical (function space) representation of the process so that there are available the shift operators θ_t defined by $X_s(\theta_t\omega) = X_{s+t}(\omega)$ all $s \geqq 0$; and the killing operators k_t defined by $X_s(k_t\omega) = X_s(\omega)$ if $s < t$, $= \Delta$ if $s \geqq t$. Further notations and terminology from general Markov theory will follow that of Blumenthal and Getoor **[2]**.

A *random time* R is a nonnegative random variable on $(\Omega, \mathscr{F})$. The *post R process* is the process $\{Y_t, t \geqq 0\}$ defined by

$$Y_t = X(R + t), \qquad t \geqq 0.$$

The *pre R process* $\{W_t, t > 0\}$ is defined by

$$\begin{aligned} W_t &= X_t, \quad t < R, \\ &= \Delta, \quad t \geqq R. \end{aligned}$$

There are a number of candidates for what should be meant by the sigma field of "events occurring before R". We shall give here only the two which figure in the subsequent exposition; others can be found in **[33]** for example. The first sigma field associated with R is

$$\mathscr{F}(R+) = \{A \in \mathscr{F}: \text{for all } t > 0, \text{ there exists } A_t \in \mathscr{F}_t \text{ such that } A \cap \{R < t\} = A_t \cap \{R < t\}\}.$$

This sigma field was first introduced in this generality in **[40]**. It is easy to see that

(a) If $R \leqq T$, then $\mathscr{F}(R+) \subset \mathscr{F}(T+)$.

(b) If $R_n \downarrow R$, then $\mathscr{F}(R+) = \bigcap \mathscr{F}(R_n+)$; in particular $\{\mathscr{F}(R+t+),\ t > 0\}$ is right continuous.

(c) $X(R)$ is $\mathscr{F}(R+)$-measurable if and only if for every t, $X(R)$ is equal on $\{R < t\}$ to some $\mathscr{F}_t$-measurable random variable.

The second sigma field associated with R is $\mathscr{F}(R)$, the sigma field generated by all events of the form $A \cap \{R > t\}$ with $A \in \mathscr{F}_t$, $t > 0$, together with the events $A \cap \{R = T\}$ with T a stopping time and $A \in \mathscr{F}_T$. This sigma field was first introduced by Pittinger and Shih **[33]**. It is easy to see that a random variable A is $\mathscr{F}(R)$-measurable if and only if there exists a well-measurable process V such that $Z = V_R$ on $\{R < \infty\}$; on the other hand Z is $\mathscr{F}(R+)$-measurable if and only if this representation holds for progressively measurable V; see **[7]**, **[25]** for the definitions of well and progressively measurable.

A fundamental problem in the theory of Markov processes is to identify the random times R for which at least one of the properties below holds:

(a) the post R process has a Markovian structure,

(b) conditional on a few appropriate random variables, the post R process is independent of $\mathscr{F}(R)$ or $\mathscr{F}(R+)$,

(c) the pre R process has a Markovian structure. When at least one of these holds, we will speak of a *path decomposition at R*. We turn now to examine aspects of several of the better known decompositions.

(2.1) *Decompositions at stopping times.* The strong Markov property holds for a right continuous Markov process $\{X_t\}$ if for every stopping time R the post R process is Markov with the same transition functions as the original process, and, conditional on $X(R)$, the post R process is independent of $\mathscr{F}(R+)$. General conditions guaranteeing the strong Markov property were first researched by Hunt **[14]**, Blumenthal **[1]**, Dynkin **[11]**, among others, and it is now well understood that under broad conditions a right continuous Markov process is strong Markov if and only if $t \to P_s f(X_t)$ is right continuous for an appropriate class of f (see **[37]** for a precise statement and further references). On the other hand, if $\{X_t\}$ is strong Markov (even a Hunt process), it is not true in general that the pre R process is Markov for all stopping times. In the mid 1950's, Hunt **[15]**, Dynkin **[10]** and others noticed that if R is a hitting time of a set A

(2.2) $$R = \inf\{t > 0 : X_t \in A\},$$

then under broad conditions **[10]**, the pre R process is Markov with transition functions Q_t given by

(2.3) $$Q_t(x, dy) = P^x\{X_t \in dy,\ t < R\}.$$

The property of hitting times that guarantees the pre R process Markov is that

(2.4) $$R = t + R \circ \theta_t \quad \text{on } \{R > t\};$$

a random time with property (2.4) is called a (perfect) *terminal time*. The chief examples of such times are of the form

(2.5) $$R = \inf\{t > 0: f(X_{t-}, X_t) \in B\}$$

where B is a Borel subset of the real line, and f a real Borel function on $E \times E$. In particular, the first jump of a given size, the first time X_{t-} enters a given set, and so forth, are terminal times. Under broad conditions, all of the terminal times are of the form (2.5) (see **[39]**) but it is technically convenient to have the abstract definition (2.4).

(2.6) *Decompositions at last exit times*. If a strong Markov process $\{X_t\}$ has a life time ζ which is finite a.s., then it often turns out (see **[6]**, **[21]**, **[32]**, for example) that the reversed process $X(\zeta - t+)$ is Markov with stationary transitions, and (less often) even strong Markov. In this case, the decompositions of paragraph (2.1) show that decomposition theorems hold for (say) hitting times defined in terms of $X(\zeta - t+)$. But (roughly speaking) a first hitting time of A for the reversed process is the time that the original process leaves A for the last time, so it is natural to expect that decomposition theorems hold at "last exit times" R defined by

(2.7) $$R = \sup\{t: X_t \in A\} \quad (= 0 \text{ if } X_t \notin A \text{ for all } t \geqq 0).$$

The property of last exit times that permit decomposition theorems to be proved can be abstracted as follows **[32]**, **[40]**: R is a *coterminal time* if

(2.8) $$\begin{aligned} &(1)\ R \circ \theta_s = (R - s)^+ && \text{for every } s,\\ &(2)\ R \circ k_s = R && \text{on } \{R < s\},\\ &(3)\ R \circ k_s \leqq s && \text{for every } s, \end{aligned}$$

As in the case of terminal times, the typical coterminal time is of the form

(2.9) $$R = \sup\{t: f(X_{t-}, X_t) \in A\} \quad (= 0 \text{ if the set of } t \text{ is empty}).$$

Decompositions at last exit times for the special case of Markov chains go back to the early 1960's (Chung **[3]**). The Markovian character of the post R process, R coterminal, was first proved in generality by Meyer-Smythe-Walsh **[40]**; see §3 for a more general formulation and explicit description of the post R process. Conditional independence was first proved by Pittinger and Shih **[33]**; see **[13]** for another approach. As shown in **[40]**, the pre R process is also Markov:

(2.10) If R is a random time satisfying (2.8), property (3), then the pre R process is Markov with transitions

$$\begin{aligned} Q_t(x, dy) &= P_t(x, dy)P^y\{R > 0\}/P^x\{R > 0\} \quad \text{if } P^x\{R > 0\} > 0,\\ &= \text{unit mass at } \Delta \quad \text{if } P^x\{R > 0\} = 0. \end{aligned}$$

The fundamental importance of hitting times to sample function analysis and to the construction of probabilistic potential theory is of course well known. Last exit times turn out to be equally important. Hunt **[16]**, followed by Nagasawa **[32]**, first noticed their key role in the theory of time reversal; their importance in probabilistic potential theory was demonstrated by (among many contributors) Kunita-Watanabe **[20]**, Port-Stone **[36]**, Chung **[5]**, Getoor-Sharpe **[12]**. In view of the duality (via time reversal) between first hits and last exits, it is perhaps not surprising that both classes of random times should be important. The duality between last exit, first hit was extended to terminal and coterminal times by Meyer-Smythe-Walsh **[40]**. Roughly speaking, to a terminal time T one may associate a coterminal time L_T by $L_T = \sup\{t: T \circ \theta_t < \infty\}$ and to a coterminal time L one may associate a terminal time T_L by $T_L = \inf\{t > 0: L(k_t\omega) > 0\}$; this relation preserves the spirit of the relation between last exits and first hits, see **[40]** for the precise result and further details. The existence of this relationship will help us understand the nature of the random times introduced in §3.

(2.11) *Randomized coterminal times, examples.* In §3 I will delineate a broad class of random times for which a decomposition theorem can be proved. Since this definition is at first sight a bit complex, I will give here a number of interesting examples, all of which have the structure exhibited in §3.

(2.12) EXAMPLE. *Minima.* Let $\{X_t\}$ be a real process and let R be the time at which it assumes its (last) ultimate minimum:

(2.13) $$R = \sup\{t: X_t \leqq X_s \text{ for all } s \leqq t\}.$$

The importance of this random time was first noticed by Williams **[41]**, **[42]**, who, in the case of diffusions, proved a decomposition theorem at R and used it as a key step in his investigation of diffusion local time; further decompositions at minima and closely related times were subsequently proved by Jacobsen **[17]** (diffusions), Pitman **[34]** (conditioned Brownian motion) and Millar **[29]** (general Markov process). If f is a real Borel function on $E \times E$, one can define in the same spirit

(2.14) $$R = \sup\{t: f(X_{t-}, X_t) \leqq f(X_{s-}, X_s), s \leqq t\}$$

and, of course, a similar definition can be made for maxima. The random times (2.14) include for example the time at which a real process makes its biggest jump, or its biggest positive jump, and so forth. A particularly interesting instance of (2.14) was discussed by Pitman **[34]**: here X_t is the h-transform of Brownian motion under a superharmonic function h, and $f = h$.

(2.15) EXAMPLE. *Random dilation.* Let the state space E be k-dimensional Euclidean space, and suppose the process starts at a point x different from 0. Let A be a closed subset of E that contains 0 but not x; assume A has nonempty interior. For each path $t \to X_t(\omega)$ expand the set A until it just touches the path. Define R to be the last t for which $X_t(\omega)$ belongs to this dilation of A. Of course, the amount A must be dilated depends on the path under consideration.

(2.16) EXAMPLE. *Last exit from a set chosen at random.* If a Borel set $A \in \mathscr{E}$ is chosen at random from a collection $\mathscr{C} \subset \mathscr{E}$, then R here will be the last time the process leaves A. Not all random choices will fit into the scheme of §3, however; it must be the case that the randomness by which A is picked is based on events before R. Here are examples; the precise interpretation of the preceding sentence is in the definition of §3. Let $\{X_t\}$ be a real process, $\lim_{t \to \infty} X_t = +\infty$; assume X starts from 0. Let M be the time of the (last) ultimate minimum (see (2.13)), let $I = \inf X_t$

be the value of this minimum, and let $T = \inf\{t: X_t < 1\}$. Here are the examples:

$$R_1 = \sup\{t: X_t \in [I, I + 1]\}, \qquad R_2 = \sup\{t: X_t \in (-\infty, M]\},$$
$$R_3 = \sup\{t: X_t \in (-\infty, T + 1]\}, \qquad R_4 = \sup\{t > M: |X_t - X_{t-}| > M\}.$$

3. Randomized coterminal times. The time of last exit from a set, the time that the ultimate minimum is achieved, and the random times listed at the end of the preceding section all have the same structure. Suppose given

(3.1a) a measure space $(A, \mathfrak{A})$,

(3.1b) a family of terminal times indexed by A, $\{T_a, a \in A\}$ such that $(a, \omega) \to T_a(\omega)$ is $\mathscr{F}$-measurable,

(3.1c) a measurable mapping Z from $(\Omega, \mathscr{F})$ to $(A, \mathfrak{A})$.

A random time R is a *randomized coterminal time* (rct) based on $(A, \mathfrak{A})$, $\{T_a\}$, Z if

(3.2a) for each $t \geqq 0$ there is an $\mathscr{F}_t$-measurable A-valued random variable Z_t such that $Z = Z_t$ on the set $\{R \leqq t\}$,

(3.2b) for each $0 \leqq s < t$ there exists $B(s, t) \in \mathscr{F}_t$ such that

$$\{s \leqq R < t\} = B(s, t) \cap \{T_{Z(\omega)}(\theta_t\omega) = +\infty\}.$$

By (3.2a) the Z in (3.2b) can be replaced by Z_t; property (3.2a) says that Z is $\mathscr{F}(R+)$-measurable.

To see the meaning of this definition in a specific instance, consider the case where $\{X_t\}$ is a real process, $\lim X_t = +\infty$, and R is the time of the (last) minimum (see (2.13)). Here $A = (-\infty, \infty)$, $\mathfrak{A} =$ Borel sigma field, $T_a = \inf\{t > 0: X_t \leqq a$ or $X_{t-} \leqq a\}$, $Z = \inf_t X_t$, and $Z_t = \inf_{s \leqq t} X_s$. Then property (3.2a) is immediate: if the ultimate minimum occurs before time t, then it is equal to the minimum attained by X on $[0, t]$. To see that (3.2b) holds, note that

$$\begin{aligned}\{s < R \leqq t\} &= \{\text{the minimum of } X \text{ occurs in } (s, t]\}\\ &= \{X \text{ goes at least as low in } (s, t] \text{ as it did before time}\\ &\qquad s, \text{ and never after } t \text{ goes as low as it did during } (s, t]\}\\ &= \{Z_{st} \leqq Z_t\}\{T_{Z_t(\omega)}(\theta_t\omega) = +\infty\},\end{aligned}$$

where $Z_{st} = \inf_{s<u\leqq t} X_u$, so $B(s, t) = \{Z_{st} \leqq Z_t]$ here. The examples (2.15), (2.16) can be checked as easily.

It is interesting to note that if R is a rct, and we then apply another rct R_1 to the post R process, then $R + R_1$ is a new rct for the original process.

(3.3) EXAMPLE. *Successive minima-maxima.* Let $\{X_t\}$ be a real process with finite lifetime. Let R be the time of the (last) ultimate minimum of X. Let R_1 be the time of the ultimate maximum of the post R process. Then $R + R_1$ is a rct. Indeed, $A = E \times E$ ($E =$ real line), $T_{x,y} = \inf\{t: X_t$ or $X_{t-} \notin (x, y)\}$ if $x < y$, $Z = (\inf_s X_s, \sup_t X_{R+t})$ and it is not hard to check properties (3.2a,b). Notice that in many cases (e.g., Brownian motion killed at an independent exponentially distributed stopping time) one can continue to apply taking maxima and minima in alternating order to the post processes, each time getting a new rct, and never in a finite number of operations reaching ζ.

According to a result of **[40]** discussed in §2, there is a coterminal time L associated with a terminal time T. Let R be a rct based on $(A, \mathfrak{A})$, $\{T_a\}$, Z; for each $a \in A$, let L_a be the coterminal time associated with T_a. The random time R then has the representation $R = L_Z$, and it is this relation that suggested the terminology. Since a coterminal time is typically a last exit time, the typical rct can be interpreted as the last leaving time of a set chosen at random (although the formal definition places a good deal of restriction on how the random set is to be picked). To illustrate, the time of the minimum (trivially) is the last time the process leaves the interval $(-\infty, \inf_s X_s]$.

In order to describe the character of the post R process when R is a rct, it is convenient to introduce the notion of "conditioned" process. If T is a terminal time, and if $P^x\{T = \infty\} > 0$, then the transitions of the original process conditioned on $\{T = \infty\}$ are again temporally homogeneous transitions:

$$\begin{aligned} P^x\{X_t \in dy \mid T = +\infty\} &= P^x\{X_t \in dy, T = \infty\}/P^x\{T = \infty\} \\ &= P^x\{X_t \in dy, t < T, T \circ \theta_t = \infty\}/P^x\{T = \infty\} \\ &= P^x\{X_t \in dy, t < T\}P^y\{T = \infty\}/P^x\{T = \infty\} \\ &= Q_t(x, dy)h(y)/h(x), \end{aligned}$$

where $Q_t(x, dy) = P^x\{X_t \in dy, t < T\}$ is the transition function of the process X killed at T (see (2.3)) and $h(y)$ is the (excessive) function $h(y) = P^y\{T = \infty\}$. Conditioned processes in the sense just defined are therefore special cases of the h-transform first introduced by Doob **[9]**.

The import of the main theorem of this section is that if R is a rct, then conditional on $Z = z$, the post R process is "just" the original process conditioned on $\{T_z = +\infty\}$. For example, if R is the time of the minimum as in (2.13), then conditional on the value of the minimum being m, the post R process is X "conditioned to stay above m forever"; i.e., conditioned on $\{T_m = \infty\}$ where $T_m = \inf\{t: X_{t-} \leqq m \text{ or } X_t \leqq m\}$. In the case of successive minimum-maximum (example (3.3)), conditional on the ultimate minimum being m_1 and the maximum after that being m_2, the post R process is just X conditioned to remain in (m_1, m_2) forever. If R is the last exit from the fixed set A, then the post R process is X conditioned to avoid A forever. We state these results formally.

(3.4) THEOREM. *Let $\{X_t, t \geqq 0\}$ be a Hunt process, and R a rct based on A, Z, $\{T_a\}$. Then for bounded Borel f,*

$$E\{f(X(R + t))|\mathscr{F}(R + s+)\} = H_{t-s}(Z; X(R + s), f), \quad 0 < s < t,$$

where $H_t(z; x, dy) = P^x\{X_t \in dy, t < T_z\}P^y\{T_z = \infty\}/P^x\{T_z = \infty\}$.

In case R is a coterminal time or the time of the minimum, this is Theorem 5.1 of **[40]** and Proposition 4.1 of **[29]** respectively. In outline, the basic idea of the proof is to introduce a discretization R_n of R by $R_n = (k + 1)/2^n$ on $\{k/2^n \leqq R < (k + 1)/2^n\}$; use property (3.2b) to convert statements about $\{k/2^n \leqq R < (k + 1)/2^n\}$ to statements involving stopping times and shifts—in a form where the Markov property can be applied, getting the desired conclusion for R_n in place of R. The proof is completed by letting $n \to \infty$ and showing that the result for R_n converges to the desired result for R; this is the most involved part of the proof, see **[29]** for the case of minima.

4. Zero-one laws and the post R process. If a decomposition of X into pre R and post R fragments is to yield insight into the functioning of the original process, it is evidently necessary to understand the stochastic structure of these fragments as processes in their own right. Often the post R process as a process in its own right fails to have some of the regularity of the original process. In this section we investigate one aspect of this problem.

Define for any random time R the sigma field $\mathscr{F}_{[R]}$ by

$$\mathscr{F}_{[R]} = \bigcap_{t>0} \sigma\{X(R+s), s<t\}. \tag{4.1}$$

If R is a stopping time, the Blumenthal zero-one law **[1]** asserts that conditional on $R < \infty$,

$$\mathscr{F}_{[R]} = \sigma\{X(R)\} \quad \text{a.s.} \tag{4.2}$$

In analogy with this, say that a zero-one law holds for a rct R based on $A, Z, \{T_a\}$ if, conditional on $R < \infty$, (4.2) holds. In contrast to the state of affairs for stopping times, zero-one laws rarely hold at a rct, thus giving one indication of how post R processes are often more perverse than the process from which they were derived. Our interest in zero-one laws is two-fold: (1) knowledge of a zero-one law at a given rct is a valuable tool in the sample function analysis of the original process; see **[28]**, **[30]**, **[31]**, for example; (2) whenever such a law can be established, it implies a strong decomposition at R: the pre and post R processes are conditionally independent, given X_R and Z. This technique of proving strong decomposition theorems was first used in **[29]**.

Let us begin by getting a clear idea how badly zero-one laws can fail. Let $\{X_t, t \geqq 0\}$ be a real process with stationary independent increments. As is well known,

$$E\exp(iuX_t) = \exp\{t\psi(u)\}, \quad \text{where} \tag{4.3}$$

$$\psi(u) = iau - (\sigma^2/2)u^2 + \int \{e^{iux} - 1 - iux(1+x^2)^{-1}\}\nu(dx). \tag{4.4}$$

Suppose $\sigma^2 > 0$ and $\nu\{(0,\infty)\} > 0$, so the process has a Brownian component, and there are jumps in the upward direction. It is well known **[19]** that $P^x\{T_y < \infty\} > 0$ where $T_y = \inf\{t > 0: X_t = y\}$; assume for convenience that $\lim_{t\to\infty} X_t = +\infty$, and define $R = \sup\{t: X_t = 0\}$. It is easy to see that $X_R = 0$, $P^0\{0 < R < \infty\} = 1$ and that X is continuous at R. Nevertheless, it may be deduced from the results in **[27]** and time reversal that both of the sets A_1, A_2 below belong to $\mathscr{F}_{[R]}$, yet have positive probability under P^0:

$$\begin{aligned} A_1 &= \{\omega: X(R+t) > 0 \text{ for all sufficiently small } t > 0\}, \\ A_2 &= \{\omega: X(R+t) < 0 \text{ for all sufficiently small } t > 0\}. \end{aligned} \tag{4.5}$$

So $\mathscr{F}_{[R]}$ is not trivial, whereas $\sigma\{X_R\}$ is and a zero-one law fails. Notice that the zero-one law at R fails despite continuity at R; intuitively, the process can leave 0 for the last time in "more than one way" and this phenomenon vitiates the zero-one property. Despite the abysmal failure of zero-one laws for the general rct, for some processes and some rct a zero-one law in fact holds. Perhaps the most interesting case so far discovered is **[29]**

(4.6) PROPOSITION. *Let $\{X_t, t \geqq 0\}$ be a real Levy process with $\nu((-\infty, \infty)) =$*

$+\infty$ *and paths that are not everywhere decreasing. Let* U *be a random time, and* R *the time point in* $[0, U]$ *at which the minimum is achieved. Assume that* $P\{R < U\} > 0$. *Then conditional on* $R < U$, $\mathscr{F}'_{[R]}$ *is trivial.*

Here $\mathscr{F}'_{[R]} = \bigcap_{s>0} \sigma\{X_{R+t} - X_R,\ t < s\}$, so the zero-one property asserted here is much stronger than that of (4.2). Here is the basic idea of the proof; consult **[29]** for the complete details which are lengthy. For simplicity, let U be a stopping time independent of $\mathscr{F}$, with an exponential distribution, and assume $P\{R < U\} = 1$. The first step is to prove that there is really only one way that a real Levy process achieves its minimum on $[0, U]$: if $Z = \inf_{t<U} X_t$ then either

$$P\{Z = X_R < X_{R-}\} = 1 \quad \text{or} \quad P\{Z = X_{R-} < X_R\} = 1 \quad \text{or} \quad P\{X_R = X_{R-}\} = 1. \tag{4.7}$$

Because the Blumenthal zero-one law holds at jump times, it is relatively easy to prove a zero-one law at R in the first two cases of (4.7). To treat the third, which is the most difficult, the next step is to show that if $\varepsilon > 0$ and R_ε is the time at which $\{X(R + t),\ t > \varepsilon\}$ achieves its minimum, then $\{X(R_\varepsilon + t) - X(R_\varepsilon),\ t \geqq 0\}$ is independent of $\mathscr{F}(R+)$ and hence independent of $\mathscr{F}'_{[R]}$. The proof of the claimed independence is a bit involved; it can be proved by checking that an analogous statement holds for random walk, and then establishing the statement in question via a limiting procedure. Once this independence is known, the triviality of $\mathscr{F}'_{[R]}$ follows at once on letting $\varepsilon \downarrow 0$, since in the case under discussion, $R_\varepsilon \downarrow R$.

Here are some other examples of rct that admit a zero-one law in the sense of (4.2).

(4.8) EXAMPLE. *Random times at local minima.* Let $\{X_t\}$ be a process whose paths agree with those of a Levy process on $[0, \zeta)$, and let R be a rct. Suppose that $P\{R \text{ is the time of a local maximum}\} = P\{R < \infty\}$. Then Proposition 4.6 can be used to show that a zero-one law holds at R. For example, if R_1 is the time of the ultimate minimum, and if R is the time of the maximum of the post R_1 process (see example (3.3)), then a zero-one law holds at R.

(4.9) EXAMPLE. *Last exit from an interval.* If $\{X_t\}$ is a real Levy process, $\lim X_t = +\infty$, $\nu((-\infty, \infty)) = \infty$, and R is the time of last exit from $(-\infty, a]$, then $\mathscr{F}_{[R]} = \sigma\{X_R\}$. The proof is similar in spirit to that of Proposition 4.4; see **[29]** for details. It turns out that $\mathscr{F}_{[R]}$ is trivial if and only if $P\{X_t \leqq X_{t-} \text{ all } t\} = 1$. It is possible to show that if $P\{X_{R-} < a < X_R\} = 1$, then $\mathscr{F}'_{[R]}$ is trivial; however, it happens in many processes that this probability is less than 1 (so also $1 > P\{X_R = X_{R-} = a\} > 0$). In this situation, $\mathscr{F}'_{[R]}$ is not trivial and, unlike the behavior at the minimum, the processes can leave $(-\infty, a]$ in "more than one way" (some paths jump out of $(-\infty, a]$ and some leave "continuously"). Fortunately, knowledge of X_R enables us to see which manner of exit has occurred. Explicit criteria in terms of ψ for the various behaviours detailed here can be found in **[29]**.

(4.10) EXAMPLE. *rct at jump times.* If R is a rct for $\{X_t\}$ such that $P\{X_R = X_{R-}\} = 0$, then since R always occurs at jump times the Blumenthal zero-one law at jump times suggests, and it can be proved that, a zero-one law holds at R. An example of this is R equal to the time at which X experiences its (last) maximum jump.

(4.11) EXAMPLE. *Real processes with no upward jumps.* Let R be a rct for a real process $\{X_t\}$ and suppose that $P\{X_t \leqq X_{t-} \text{ all } t\} = 1$. Suppose that

$P\{X_{R+t} > X_R \text{ for all } t > 0\} = 1$. Then a zero-one law holds at R. To illustrate, let $\{X_t\}$ be a real diffusion with $\lim X_t = \infty$. Then all the terminal times are hitting times of sets; from this, any rct R must then be the last exit from some set chosen at random, so one sees that in this case a zero-one law holds for every rct.

5. Conditional independence. While the Markovian nature of the post R process can be proved for any rct, establishing conditional independence in general seems several orders of magnitude more difficult. The relation desired is

$$(5.1)\quad \begin{aligned} &E\{f(X(t_1 + R), \cdots, X(t_n + R)) \,|\, \mathscr{F}(R+)\} \\ &\qquad = E\{f(X(t_1 + R), \cdots, X(t_n + R)) \,|\, X_R, Z\} \end{aligned}$$

which asserts that the evolution of the post R process depends on the past only through $X(R)$ and Z. Because of (3.4) it is necessary to establish only

$$(5.2)\quad E\{f(X(t + R)) \,|\, \mathscr{F}(R+)\} = E\{f(X(t + R)) \,|\, X_R, Z\}.$$

It then follows that the left side of (5.1) is

$$(5.3)\quad \int f(x_1, \cdots, x_n)\, H_{t_n - t_{n-1}}(Z; x_{n-1}, dx_n) \cdots H_{t_2 - t_1}(Z; x_1, dx_2)\, Q_{t_1}(Z, X(R); dx_1),$$

where $Q_s(Z, X(R); dy)$ is a regular conditional distribution of $X(R + s)$ given $(Z, X(R))$.

If a zero-one law holds at R, then (5.2) is immediate; indeed, by (3.4), a martingale convergence theorem, and the right continuity of $\{\mathscr{F}(R + t+)\}$,

$$\begin{aligned} E\{f(X(R + t)) \,|\, \mathscr{F}(R+)\} &= \lim_{s \downarrow 0} E\{f(X(R + t)) \,|\, \mathscr{F}(R + s+)\} \\ &= \lim_{s \downarrow 0} H_{t-s}(Z; X(R + s), f). \end{aligned}$$

The presumed zero-one law forces this last limit to belong to $\sigma\{X(R), Z\}$, yielding (5.2). Combining the observation just made with the results of §4 gives several strong decomposition theorems.

(5.4) PROPOSITION. *Let $\{X_t, t > 0\}$ be a strong Markov process with lifetime ζ whose paths agree with those of a real Levy process on* $[0, \zeta)$. *Let R be a rct such that $P\{R \text{ is the time of a local minimum}\} = P\{R < \infty\}$. Then conditional on Z and X_R, the post R process is independent of $\mathscr{F}(R+)$, and it is Markov with transitions $H_t(X(R), Z; dy)$ and entrance law $Q_s(X(R), Z; dy)$ given in* (3.4) *and* (5.3) *respectively.*

Other R for which conditional independence of $\mathscr{F}(R+)$ and the post R can be proved via zero-one laws include:

(a) those R that always occur at a jump time, such as the time of the maximum jump of a process;

(b) certain R for processes with no upward jumps (see (4.11)); in particular, any rct for a real diffusion with $\lim X_t = \infty$.

As we have seen, however, there are many rct for which zero-one laws fail. In some of these cases (e.g., the example around (4.5)) conditional independence can still be proved by probing deeper the structure of the sigma field $\mathscr{F}_{[R]}$. The only other broad class of rct's for which conditional independence has been proved is the class of coterminal times. Here is (a special case of) the basic theorem first proved by Pittenger-Shih [**33**]; other approaches are due to Getoor-Sharpe [**13**] and Maisonneuve-Meyer [**23**].

(5.5) PROPOSITION. *Let $\{X_t\}$ be a Hunt process, and let R be a coterminal time. Then the post R process is conditionally independent of $\mathscr{F}(R)$, given X_R.*

In brief, when the rct R based on A, Z, $\{T_a\}$ in question is not randomized (i.e., A consists of a single point), conditional independence can be proved, at least for $\mathscr{F}(R)$, if not $\mathscr{F}(R+)$. The randomization of the problem involved in taking more complicated A seems to destroy the homogeneity essential to the methods of [**33**], [**13**], [**23**] so these methods apparently do not extend to other rct.

The general problem of characterizing *all* random times R for which the post R process is (conditionally) Markovian and/or for which the post R process is conditionally independent of $\mathscr{F}(R)$ is very far from solution.In the case of discrete time, discrete state Markov processes, Jacobsen and Pitman [**18**] have succeeded in making some interesting progress:

(5.6) PROPOSITION. *If $\{X_n, n = 1, 2, \cdots\}$ is a Markov chain with discrete state space E, if R is a random time such that*

(a) *the post R process is Markov with a fixed transition matrix q,*

(b) *the post R and pre R processes are conditionally independent given X_R.*

Then $R = R_1 + R_2$ where R_1 is a coterminal time for $\{X_n\}$ and R_2 is a stopping time for the sigma fields $\{\mathscr{F}(R_1 + n)\}$.

6. Path decompositions, an example of Williams. In this section we illustrate the results discussed above with a decomposition result for Levy processes. In the case of Brownian motion, this decomposition was proved with a bare hands approach by Williams [**42**]; further discussion and closely related results in the Brownian case are given by Jacobsen [**17**], and Pitman [**35**]. As remarked before, this decomposition was a key step in Williams deep investigation of Brownian local time [**42**], [**24**]. Let $\{X_t\}$ be a real Levy process with $\lim\sup_{t\to\infty} X_t = +\infty$. Suppose that $P^x\{T_y < \infty\} > 0$ for all x, y where $T_y = \inf\{t > 0: X_t = y\}$; the precise criteria for this were established by Kesten [**19**]. Let $T = \inf\{t > 0: X_t > 1\}$. Then the killed process

$$\begin{aligned} X_t^1 &= X_t, \qquad t < T, \\ &= \Delta, \qquad t \geqq T, \end{aligned}$$

is a strong Markov process with transitions $K_t(x, dy)$ given by (2.3):

$$K_t(x, dy) = P^x\{X_t \in dy,\ t < T\}.$$

Let L be the last exit time of the process X^1 from the point $\{0\}$:

$$L = \sup\{t: X_t^1 = 0\}.$$

Then $X_L^1 = 0$, and since L is a coterminal time for X^1, it follows from (5.5) that the pre L and post L parts of X^1 are independent. By (3.4), the post L part of X^1 is strong Markov with transitions

$$H_t(x, dy) = P^x\{X_t \in dy,\ t < T \wedge T_0\}P^y\{T_0 < T\}/P^x\{T_0 < T\}, \qquad x < 1.$$

These transitions can be computed exactly in the Brownian case. On the other hand, by (2.10) the pre L part of X^1 is also strong Markov, its transitions $G_t(x, dy)$ being given by

$$G_t(x, dy) = P^x\{X_t \in dy, t < T\}P^y\{T_0 < T\}/P^x\{T_0 < T\}, \qquad x > 1.$$

The paths of the pre L part $\{Y_t\}$ of X^1 obviously coincide with those of a Levy process up to the lifetime of $\{Y_t\}$. Therefore by (4.6), if R is the time at which $\{Y_t\}$ assumes its maximum, the pre R and post R parts of Y are conditionally independent, given X_R and $Z = \sup Y_t$. The transition function of the post R part of Y can then be determined via (4.6).

References

1. R. M. Blumenthal, *An extended Markov property*, Trans. Amer. Math. Soc. **85** (1957), 52–72. MR **19**, 468.

2. R. N. Blumenthal and R. K. Getoor, *Markov processes and potential theory*, Academic Press, New York and London, 1968. MR **41** #9348.

3. K. L. Chung, *On last exit times*, Illinois J. Math. **4** (1960), 629–639. MR **24** #A575.

4. ———, *Markov chains with stationary transition probabilities*, 2nd ed., Springer-Verlag New York, 1967. MR **36** #961.

5. ———, *Probabilistic approach to the equilibrium problem in potential theory*, Ann. Inst. Fourier (Grenoble).

6. K. L. Chung and J. B. Walsh, *To reverse a Markov process*, Acta Math. **123** (1969), 225–251. MR **41** #2761.

7. C. Dellacherie, *Capàcités et processus stochastiques*, Ergebnisse d. Math. 67, Springer-Verlag, Berlin, 1972.

8. J. L. Doob, *Markoff chains—denumerable case*, Trans. Amer. Math. Soc. **58** (1945), 455–473. MR **7**, 210.

9. ———, *Conditional Brownian motion and the boundary limits of harmonic functions*, Bull. Soc. Math. France **85** (1957), 431–458. MR **22** #844.

10. E. B. Dynkin, *Markov processes*. I, II, Academic Press, New York, 1965. MR **33** #1887.

11. E. B. Dynkin and A. A. Yushkevich, *Strong Markov processes*, Theor. Probability Appl. **1** (1956), 134–139.

12. R. K. Getoor and M. J. Sharpe, *Last exit times and additive functionals*, Ann. Probability **1** (1973), 550–569.

13. ———, *Last exit decompositions and distributions*, Indiana Univ. Math. J. **23** (1973/74), 377–404. MR **48** #12654.

14. G. A. Hunt, *Some theorems concerning Brownian motion*, Trans. Amer. Math. Soc. **81** (1956), 294–319. MR **18**, 77.

15. ———, *Markoff processes and potentials*. I, II, III, Illinois J. Math. **1** (1957), 44–93, 316–369; **2** (1958), 151–213. MR **19**, 951; **21** #5824.

16. ———, *Markoff chains and Martin boundaries*, Illinois J. Math. **4** (1960), 313–340. MR **23** #A691.

17. M. Jacobsen, *Splitting times for Markov processes and a generalized Markov property for diffusions*, Z. Wahrscheinlichkeitstheorie und Verw. Gebiete **30** (1974), 27–43. MR **51** #11670.

18. M. Jacobsen and J. Pitman, *Birth, death, and conditioning of Markov chains* (to appear).

19. H. Kesten, *Hitting probabilities of single points for processes with stationary independent increments*, Mem. Amer. Math. Soc., No. **93** (1969). MR **42**, 6940.

20. H. Kunita and T. Watanabe, *Markov processes and Martin boundaries*. I, Illinois J. Math. **9** (1965), 485–526. MR **31** #5240.

21. ———, *On certain reversed processes and their applications to potential theory and boundary theory*, J. Math. Mech. **15** (1966), 393–434. MR **33** #1891.

22. B. Maisonneuve, *Exit systems*, Ann. Probability **3** (1975), 399–411.

23. B. Maisonneuve and P.-A. Meyer, *Ensembles aleatoires markoviens homogenes.* I, II, III, IV, Séminaire de Probabilities, Strasbourg VIII. Springer Lecture Notes (1973).

24. H. P. McKean, *Brownian local times*, Advances in Math. **16** (1975), 91–111. MR **51** #7018.

25. P.-A. Meyer, *Probability and potentials*, Blaisdell, Waltham, Mass., 1966. MR **34** #5119.

26. ———, *Processus de Markov*, Lecture Notes in Math., No. 26, Springer-Verlag, Berlin, New York, 1967. MR **36** #2219.

27. P. W. Millar, *Exit properties of stochastic processes with stationary independent increments*, Trans. Amer. Math. Soc. **178** (1973), 459–479. MR **47** #9731.

28. ———, *Sample functions at a last exit time*, Z. Wahrscheinlichkeitstheorie **34** (1976), 91–111.

29. ———, *Zero one laws and the minimum of a Markov process*, Trans. Amer. Math. Soc. (to appear).

30. D. Monrad, *Stable processes: sample function growth at a last exit time* (to appear).

31. ———, *Asymmetric Cauchy processes: sample functions at last zero* (to appear).

32. M. Nagasawa, *Time reversions of Markov processes*, Nagoya Math. J. **24** (1964), 177–204. MR **29** #6542.

33. A. O. Pittenger and C. T. Shih, *Coterminal families and the strong Markov property*, Trans. Amer. Math. Soc. **182** (1973), 1–42. MR **49** #1600.

34. J. Pitman, *Path decomposition for conditional Brownian motion*, Inst. Math. Statist., Univ. Copenhagen, preprint no. 11 (1974).

35. ———, *One-dimensional Brownian motion and the three-dimensional Bessel process*, J. Appl. Probability **7** (1975), 511–526. MR **51** #11677.

36. S. Port and C. Stone, *Infinitely divisible processes and their potential theory.* I, II, Ann. Inst. Fourier (Grenoble) **21** (2)(1971), 157–275; **21** (4)(1971), 179–265. MR **49** #11640.

37. John B. Walsh, *Some remarks on the Feller property*, Ann. Math. Statist. **41** (1970), 1672–1683. MR **42** #5328.

38. ———, *Time reversal and the completion of Markov processes*, Invent. Math. **10** (1970), 57–81. MR **42** #5329.

39. John B. Walsh and M. Weil, *Représentation de temps terminaux et applications aux fonctionnelles additives et aux systèmes de Lévy*, Ann. Sci. Ecole Norm Sup (4) **5** (1972), 121–155. MR **46** #2750.

40. J. B. Walsh, P.-A. Meyer and R. Smythe, *Birth and death of Markov processes*, Proc. 6th Berkeley Sympos. Math. Statist. Prob., Univ. Calif., Vol. III, 295–306 (1972).

41. David Williams, *Decomposing the Brownian path*, Bull. Amer. Math. Soc. **76** (1970), 871–873. MR **41** #2777.

42. ———, *Path decomposition and continuity of local time for one-dimensional diffusions.* I, Proc. London Math. Soc. (3) **28** (1974), 738–768. MR **50** #3373.

UNIVERSITY OF CALIFORNIA, BERKELEY

Proceedings of Symposia in Pure Mathematics
Volume 31, 1977

STOCHASTIC STABILITY AND BOUNDARY PROBLEMS

MARK A. PINSKY*

1. Introduction. Many modern developments in potential theory were inspired by the 1906 paper of Fatou [3] on the existence of boundary limits of bounded harmonic functions in the unit disc. In the case of Laplace's equation these ideas were further developed by many authors [1], [2], [8], [9]. For other nondegenerate elliptic equations, it was shown [14] that a similar theory can be erected, based on an appropriate generalization of the Harnack inequality [13].

In the case of degenerate elliptic equations, the Harnack inequality is not available, and other methods must be used to study the Dirichlet problem and Fatou-type theorems. In this regard, the methods of stochastic stability [4], [5], [7] have recently proved fruitful in isolating a class of degenerate elliptic equations which can be treated rather completely. We are still quite far from understanding the general degenerate elliptic equation; it is hoped that the methods presented here may prove to be useful in other situations.

In §2 we give a self-contained account of the necessary parts of stochastic stability. In §3 we formulate the Dirichlet problem and in §4 we formulate the Fatou-type theorems.

2. Stochastic stability. Consider a system of Itô differential equations

$$dx_i = \sigma_{ij}(x)dw_j + b_i(x)\,dt \qquad (1 \leqq i \leqq n) \tag{1}$$

where σ_{ij}, b_i are smooth and bounded functions on R^n, $1 \leqq i, j \leqq n$ (repeated indices imply a summation); (w_i) is an n-dimensional Wiener process. It is well known that the solution X_t^x of (1) with $X_0^x = x$ is a strong Markov process with continuous paths. We assume further that

---

AMS (MOS) subject classifications (1970). Primary 35J25; Secondary 35J70, 60H10, 31A10.

*Supported by National Science Foundation grant MPS71–02838 A04.

(2) $$\sigma_{ij}(0) = 0 = b_i(0).$$

In view of the uniqueness of the solution and the smooth coefficients, we have a.s. that, for $t > 0$, $X_t^x = 0$ iff $x = 0$, i.e., zero is an unattainable trap. Following Hasminskiĭ **[5]** we introduce the following concept of stability:

DEFINITION. The zero solution of (1) is *asymptotically stable in probability* (ASIP) iff

(3) $$\lim_{x\to 0} P\left\{\lim_{t\to\infty} X_t^x = 0\right\} = 1.$$

This form of stability is very close to "regularity" in potential theory. In order to verify the presence of stability, we introduce the following "Liapunov criterion".

DEFINITION. An *S-function* is a real-valued function defined on a deleted neighborhood U of $0 \in R^n$ such that

(4a) $$f \in C^2(U),$$
(4b) $$|\sigma \nabla f| \leqq \text{const} \qquad (x \in U),$$
(4c) $$Lf \leqq -1 \qquad (x \in U),$$
(4d) $$\lim_{x\to 0} f(x) = -\infty,$$

where L is the infinitesimal generator of the Itô process (1). Roughly speaking, an S-function plays the role for stochastic equations that the logarithm of the classical Liapunov function plays for systems of ordinary differential equations.

PROPOSITION. *Let f be S-function for* (1). *Then the zero solution is ASIP.*

The complete proof is given elsewhere **[10]** and is a straightforward consequence of Itô's formula. In the presence of an S-function we have exponential asymptotic stability almost surely for those paths which remain forever in a given neighborhood of zero.

EXAMPLE 1. For the general one-dimensional equation we have

(5) $$dx = [\sigma x + \varepsilon_1(x)]\, dw + [bx + \varepsilon_2(x)]\, dt,$$

where $\lim_{x\to 0} \varepsilon_i(x)/|x| = 0$. In this case we take $f(x) = \log|x|$. Then a direct calculation shows that $\lim_{x\to 0} Lf(x) = Q$, where

(6) $$Q = b - \tfrac{1}{2}\sigma^2.$$

Therefore we have ASIP whenever $Q < 0$. Of course this can also be inferred from the general theory of one-dimensional diffusion **[6]** where ASIP goes by the term "attractive unattainable boundary point".

EXAMPLE 2. For the linear equation in R^n, we have

(7) $$dx_i = \sigma_{ijk} x_j dw_k + b_{ij} x_j dt.$$

Let $a_{ij}(x) = \sigma_{ikl}\sigma_{jpl} x_k x_p$, and set

(8) $$Q(x) = b_{ij}x_i x_j/|x|^2 + \tfrac{1}{2}a_{ii}(x)/|x|^2 - a_{ij}(x)x_i x_j/|x|^4.$$

A direct calculation shows that $L(\log|x|) = Q(x)$. Thus if $Q(x) \leqq -A^2 < 0$ we clearly have ASIP. Under a supplementary hypothesis of nondegeneracy, Hasminskiĭ **[5]** showed ASIP if $\int Q(x)m(dx) < 0$ where $m(dx)$ is a Radon measure on $|x| = 1$.

EXAMPLE 3. The following example arises in the Dirichlet problem in R^2. Consider the system (1) where $n = 2$. Assume the following conditions:

(9) $$\sigma_{2j}(x_1, 0) = 0 = b_2(x_1, 0),$$

(10) $$Q_{\parallel} \equiv \lim_{x_1 \to 0} \{b_1(x_1, 0)/x_1 - \tfrac{1}{2}\sigma_{1j}^2(x_1, 0)/x_1^2\} < 0,$$

(11) $$Q_{\perp} \equiv \lim_{x_2 \to 0} \{b_2(0, x_2)/x_2 - \tfrac{1}{2}\sigma_{2j}^2(0, x_2)/x_2^2\} < 0.$$

Condition (9) signifies that $(x_2 = 0)$ is an invariant manifold for the diffusion: $(X_t^{(x_1,0)})_2 = 0$ a.s. for $t > 0$. Condition (10) is the one-dimensional stability condition of Example 1 above. Condition (11) signifies that the x_2-coordinate is ASIP (although not itself a Markov process). Under these hypotheses it was shown in **[10]** that zero is ASIP and that an S-function can be constructed in the form $f(x_1, x_2) = c \log(x_1^2 + x_2^2) + h(x_2/x_1)$ where c is a positive constant and h is an appropriate smooth function.

In this example we can study the following more refined asymptotic behavior by introducing an angular coordinate: let

$$\phi = \tan^{-1}(x_2/x_1), \qquad b(\phi) = \lim_{r \to 0}(L\phi)(r\cos\phi, r\sin\phi),$$
$$\sigma^2(\phi) = \lim_{r \to 0} L(\phi^2)(r\cos\phi, r\sin\phi) - 2\phi b(\phi).$$

Assume that

(12) $$\sigma^2(\phi) > 0, \qquad 0 < \phi < \pi.$$

PROPOSITION. *Let conditions* (9)—(12) *hold. If* $Q_{\parallel} < Q_{\perp} < 0$, *then for any* $x_2 > 0$

(13) $$P\left\{\liminf_{t \to \infty} \phi(t) = 0, \limsup_{t \to \infty} \phi(t) = \pi \,\middle|\, \lim_{t \to \infty} X_t^x = 0\right\} = 1.$$

If $Q_{\perp} < Q_{\parallel} < 0$, *then for any* $x_2 > 0$ *we have*

(14) $$\lim_{x \to 0; \phi(x) \to 0} P\left\{\lim_{t \to \infty} \phi(t) = 0\right\} = 1,$$

(15) $$\lim_{x \to 0; \phi(x) \to \pi} P\left\{\lim_{t \to \infty} \phi(t) = \pi\right\} = 1.$$

The proof is given in **[12]**.

3. Application to the Dirichlet problem. Let D be a plane domain whose boundary consists of a single simple closed curve of class C^∞. We are given a degenerate elliptic operator

(16) $$L = \tfrac{1}{2} a_{ij}\partial^2/\partial x_i \partial x_j + b_i \partial/\partial x_i$$

whose coefficients satisfy the following conditions:

(17) a_{ij}, b_i are C^∞ and bounded on some open set containing D. On every compact subset of D, a_{ij} is nondegenerate.

(18) The boundary is the locus $\{x: \psi(x) = 0\}$ where ψ is positive inside D and $\nabla\psi \neq 0$ on ∂D. On ∂D we have

$$\mathscr{A}(x) \equiv a_{ij}\psi_i\psi_j = 0, \qquad \mathscr{B}(x) \equiv b_i\psi_i + \tfrac{1}{2} a_{ij}\psi_{ij} = 0.$$

(19) On ∂D we have

$$Q_\perp(\bar{x}) \equiv \lim_{x\to\bar{x}} \left\{ \frac{\mathscr{B}(x)}{\phi(x)} - \frac{1}{2}\frac{\mathscr{A}(x)}{\phi(x)^2} \right\} < 0 \qquad (\bar{x} \in \partial D).$$

From conditions (17)—(18) we see that on ∂D there lives a smooth one-dimensional diffusion process. If the boundary curve has the equation $x_1 = x_1(\theta)$, $x_2 = x_2(\theta)$, $0 \leqq \theta \leqq 2\pi$, then we can write the "boundary operator" in the form

$$Lf(\theta) = \tfrac{1}{2}\bar{a}(\theta)f''(\theta) + \bar{b}(\theta)f'(\theta)$$

where $\bar{a}(\theta)$, $\bar{b}(\theta)$ are smooth functions. We assume in addition that $\bar{a}(\theta) + \bar{b}(\theta)^2$ has a finite number of zeros $\{\theta_1, \cdots, \theta_k\}$, $1 \leqq k < \infty$.

(20) $$Q_\parallel(\theta_i) = \lim_{\theta\to\theta_i} \{\bar{b}(\theta)/(\theta - \theta_i) - \tfrac{1}{2}\bar{a}(\theta)/(\theta-\theta_i)^2\} < 0.$$

$\bar{b}(\theta) \geqq 0$ whenever $\bar{a}(\theta) = 0$.

In the terminology of one-dimensional diffusion all traps are attractive and unattainable; all shunts have the same orientation. In the following we use θ_i to denote the point $(x_1(\theta_i), x_2(\theta_i))$.

THEOREM. *Consider the Dirichlet problem for* $Lu = 0$, *with the boundary conditions* $\lim_{x\to\theta_i} u(x) = c_i$ $(1 \leqq i \leqq k)$. *This problem has a unique* C^2 *solution, given by the formula*

$$u(x) = \sum_{i=1}^{k} c_i\, P\left\{\lim_{t\to\infty} X_t^x = \theta_i\right\}.$$

The main step in the proof [**11**] is to show that the above probabilities add up to 1 for each $x \in D$. That u satisfies the boundary condition simply reflects the fact that θ_i is ASIP.

4. Fatou-type theorems. Let L satisfy (16)—(20) in D. By locally transforming the boundary to a half-space we assume that (12) holds at θ_i, $1 \leqq i \leqq k$.

If we are given a bounded solution of $Lu = 0$ in D, can we assert the existence of $\lim u(x)$ for an arbitrary manner of (nontangential) approach to the boundary? By using the above methods we obtain the following result.

THEOREM. *Let* $u \in C^2(D)$ *be a solution of* $Lu = 0$. *Assume that in a neighborhood of* θ_i *that* $\nabla u = o(1/R)$, $\nabla^2 u = o(1/R^2)$ *where* R *is the distance to* θ_i.

(21) *If* $Q_\parallel(\theta_i) < Q_\perp(\theta_i) < 0$ *then there exists* $\lim_{x\to\theta_i} u(x)$.

(22) *If* $Q_\perp(\theta_i) < Q_\parallel(\theta_i) < 0$ *then for each* $\phi \in (0, \pi)$ *there exists* $\bar{u}(\phi) = \lim_{x\to\theta_i;\, \phi(x)\to\phi} u(x)$. *Furthermore* $\bar{u}(\phi) = c_1 + c_2\pi(\phi)$ *where* c_1, c_2 *are constants and* π *is a (nonconstant) solution of* $L\pi(\phi) = 0$, $0 < \phi < \pi$.

The proof [**12**] depends on rewriting the equation $Lu = 0$ in the form

$$\frac{1}{2}\sigma^2(\phi)\frac{\partial^2 u}{\partial\phi^2} + b(\phi)\frac{\partial u}{\partial \phi} = g,$$

where $\lim_{x\to\theta_i} g(x) = 0$.

REFERENCES

1. M. Brelot and J. L. Doob, *Limites angulaires et limites fines*, Ann. Inst. Fourier (Grenoble) **13** (1963), fasc. 2, 395–415. MR **33** #4299.

2. J. L. Doob, *Conditional Brownian motion and the boundary limits of harmonic functions*, Bull. Soc. Math. France **85** (1957), 431–458. MR **22** #844.

3. P. Fatou, *Series trigonometriques et series de Taylor*, Acta Math. **30** (1906), 335–400.

4. A. Friedman, *Stochastic differential equations*. Vols. I, II, Academic Press, New York, 1975.

5. R. Z. Has′minskiĭ, *Stability of systems of differential equations under random perturbations of their parameters*, "Nauka", Moscow, 1969. (Russian) MR **41** #3925.

6. K. Itô and H. P. McKean, Jr., *Diffusion processes and their sample paths*, Academic Press, New York; Springer-Verlag, Berlin, 1965. MR **33** #8031.

7. H. J. Kushner, *Stochastic stability and control*, Academic Press, New York and London, 1967. MR **35** #7723.

8. R. S. Martin, *Minimal positive harmonic functions*, Trans. Amer. Math. Soc. **49** (1941), 137–172. MR **2**, 292.

9. L. Naïm, *Sur le rôle de la frontière de R. S. Martin dans la théorie du potentiel*, Ann. Inst. Fourier (Grenoble) **7** (1957), 183–281. MR **20** #6608.

10. M. Pinsky, *Stochastic stability and the Dirichlet problem*, Comm. Pure Appl. Math. **27** (1974), 311–350. MR **51** #6987.

11. ———, *Asymptotic stability and angular convergence of stochastic systems*, Math. Programming Studies no. 5, North-Holland, 1976.

12. ———, *Isolated singularities of degenerate elliptic equations in* R^2, J. Differential Equations (to appear).

13. J. B. Serrin, Jr., *On the Harnack inequality for linear elliptic equations*, J. Analyse Math. **4** (1955/56), 292–308. MR **18**, 398.

14. M. G. Šur, *The Martin boundary for a linear elliptic operator of the second-order*, Izv. Akad. Nauk SSSR Ser. Mat. **27** (1963), 45–60; English transl., Amer. Math. Soc. Transl. (2) **56** (1966), 19–36. MR **27** #1690.

NORTHWESTERN UNIVERSITY

Proceedings of Symposia in Pure Mathematics
Volume 31, 1977

SOME SAMPLE PATH PROPERTIES OF THE ASYMMETRIC CAUCHY PROCESSES*

WILLIAM E. PRUITT AND S. JAMES TAYLOR

1. Main results. The Cauchy processes in $\boldsymbol{R}^1$ are processes with stationary independent increments and characteristic function

$$Ee^{iuX_t} = e^{-t|u|(1+ih \operatorname{sgn} u \log|u|)},$$

where h is an asymmetry parameter which must be in the interval $[-2/\pi, 2/\pi]$. The processes obtained from these by multiplication by a scale factor and addition of a linear drift term are also Cauchy processes. However, these changes do not affect the properties we will describe so we will simplify the notation by not bringing in these two additional parameters. If $h = 0$, the process is the symmetric Cauchy process while if $h = \pm 2/\pi$, these processes are called completely asymmetric. Except for making a few comments for purposes of comparison we will consider only the asymmetric processes. Also, since changing the sign of h has the same effect as looking at the process $-X_t$, we will assume from now on that $h > 0$. It is always possible to choose a version of the process which is strong Markov and has certain path regularity properties such as right continuity. We will assume that this has been done in order that we may use some of the general potential theory of Hunt.

Several rather surprising local features of the asymmetric Cauchy processes have been discovered in the last decade. Port and Stone [3] showed that x is regular for $\{x\}$ for these processes, i.e., they have the property of returning immediately to the starting point almost surely. This is in sharp contrast to the symmetric Cauchy process which has probability zero of hitting any given point. As a consequence, these processes have a local time but Getoor and Kesten [1] showed that it must be discontinuous. Again, this was in contrast to the processes having local times which

AMS (*MOS*) *subject classifications* (1970). Primary 60G17, 60J30.
*The preparation of this paper was supported in part by the National Science Foundation.

had been previously studied; in these earlier cases the local times were known to have rather storng continuity properties. Millar and Tran [2] then proved that the asymmetric Cauchy local time was, in fact, almost surely unbounded on every neighborhood of the starting point.

We have recently discovered some rather surprising global properties of these processes. Some of these will be described as soon as we make a few general remarks. Although the symmetric Cauchy process is interval recurrent, the asymmetric ones are transient. With the exception of the completely asymmetric process, they have both positive and negative jumps and

$$\liminf_{t\to\infty} X_t = -\infty, \qquad \limsup_{t\to\infty} X_t = +\infty.$$

The completely asymmetric process has only positive jumps although it will initially spend some time in $(-\infty, 0)$ by drifting there. It does have a minimum value, however, and $\lim_{t\to\infty} X_t = +\infty$.

Let R denote the range of the process, i.e.,

$$R = R(\omega) = \{x: X_t = x \text{ for some } t \geqq 0\}.$$

We also need to introduce a parameter $\rho = (2 - \pi h)/(2 + \pi h)$. A Cauchy process X_t with $h = 2(p-q)/\pi$ can be constructed from two independent completely asymmetric Cauchy processes Y_t and Z_t by letting $X_t = pY_t - qZ_t$ where $p > 0$, $q > 0$, $p+q = 1$. Thus $\rho = q/p$ can be thought of as the relative contribution of the negative jumps compared to the positive jumps.

THEOREM 1. *If* $|\cdot|$ *denotes Lebesgue measure, then as* $a \to \infty$

$$|R \cap [0, a]| \log a/a \xrightarrow{\mathscr{D}} G$$

where G is a geometric distribution assigning measure $(1-\rho)\rho^{j-1}$ *to* j *for* $j = 1, 2, \cdots$. *Also,*

$$|R \cap [-a, 0]| \log a / a \xrightarrow{\mathscr{D}} H$$

where H is geometric but assigns measure $(1-\rho)\rho^{j}$ *to* j *for* $j = 0, 1, \cdots$.

Thus we have the surprising result that the measure of $R \cap [0, a]$ is relatively close to an integral multiple of $a/\log a$ for large a with high probability. Our original proof of this result used Laplace transforms but we have recently found a more probabilistic proof which sheds some light on this curious fact. We will give this proof in the next section.

Note that in the completely asymmetric case $\rho = 0$ and the first statement of the theorem becomes

$$|R \cap [0, a]| \log a/a \xrightarrow{P} 1.$$

In this case it is natural to ask if the convergence is almost sure. A simple argument based on the next theorem will show that this cannot be the case. However, it is trivially true in the completely asymmetric case that $|R \cap [-a, 0]| \log a/a \xrightarrow{\text{a.s.}} 0$ since the process has a minimum.

The next theorem gives integral tests for the sizes of the large holes that occur in the range of these processes.

THEOREM 2. *Let $\varphi(x)$ be a positive increasing function. Then*

$$P\{R \cap [n, n\varphi(n)] = \varnothing \; i.o.\} = 0 \quad \textit{iff} \int^{\infty} \frac{dx}{x \log x \varphi(x)} < \infty,$$

$$= 1 \quad \textit{iff} \int^{\infty} \frac{dx}{x \log x \psi(x)} = \infty,$$

and

$$P\{R \cap [-\varphi(n), -n] = \varnothing \; i.o.\} = 0 \quad \textit{iff} \int^{\infty} \left(\frac{\log x}{\log \varphi(x)}\right)^{\rho/(1-\rho)} \frac{dx}{x \log x} < \infty,$$

$$= 1 \quad \textit{iff} \int^{\infty} \left(\frac{\log x}{\log \varphi(x)}\right)^{\rho/(1-\rho)} \frac{dx}{x \log x} = \infty.$$

The first thing to notice about this result is the vastly different sizes of the large holes in the two directions. For the positive axis a borderline φ is given by $\log \log x$ and there will be infinitely many holes in the range of the type $[n, n \log \log n]$. For the negative axis a borderline φ is

$$\varphi(x) = \exp\{\log x (\log \log x)^{(1-\rho)/\rho}\}$$

which is larger than any power of x. Thus, for example, there will be infinitely many holes in the range of the type $[-n^k, -n]$ for any k.

To return to the comment we made above about $|R \cap [0, a]|$ in the completely asymmetric case, note that by Theorem 2, for a given ω, we can find arbitrarily large values of n such that $[n, n \log \log n]$ is a hole in the range. But then

$$|R \cap [0, n]| = |R \cap [0, n \log \log n]|$$

while the normalizing factors will differ by a factor of $\log \log n$. Thus it is clearly impossible to have almost sure convergence to one.

The proof of Theorem 2 will appear in **[4]** but a comment about it may be of some interest. The critical step is to obtain a precise asymptotic estimate for the probability that the process misses an interval of the form $[n, n\varphi(n)]$ or $[-\varphi(n), -n]$. This is done by using general potential theory. The unusual feature is that sharp estimates are needed for the amount of mass the capacitary distribution on a large interval assigns to certain subintervals. For example, one such result is that if μ is the capacitary measure on $[-a, 0]$, then

$$\mu[-x, 0] \sim \left(\frac{\log x}{\log a}\right)^{1/(1-\rho)} \mu[-a, 0]$$

uniformly in x for $a \geqq x \geqq e^{(\log \log a)^5}$ as $a \to \infty$.

2. Proof of Theorem 1. We start by collecting some of the basic facts which we will need in the proof. By the Fourier inversion theorem, the density function of X_t is

$$p(t, x) = (2\pi)^{-1} \int \exp\{-iux - t|u|(1 + ih \operatorname{sgn} u \log|u|\} \, du.$$

Although this integral cannot be evaluated explicitly, it is clear that the density satisfies the scaling property

$$p(t, x) = p(1, xt^{-1} - h \log t)t^{-1}.$$

It is also possible to obtain very precisely the asymptotic behavior of $p(1, x)$. These facts about the density can then be used to obtain information about the potential kernel $g(x) = \int_0^\infty p(t, x)\, dt$. Port and Stone [**3**] showed that g is continuous and that

$$\lim_{x\to-\infty} \log|x| g(x) = c, \qquad \lim_{x\to+\infty} \log x\, g(x) = b,$$

where $c = (2 - \pi h)/2\pi h^2$, $b = (2 + \pi h)/2\pi h^2$. In the present paper we will need the sharper form of the asymptotic behavior of the kernel which we obtained in [**4**]:

(1)
$$g(x) = \frac{c}{\log |x|} + O\left(\frac{1}{\log^2 |x|}\right), \qquad x \to -\infty,$$
$$g(x) = \frac{b}{\log x} + O\left(\frac{1}{\log^2 x}\right), \quad x \to +\infty.$$

For a compact set E, there is a measure μ on E called the capacitary distribution which has the property that, for all $x \in \boldsymbol{R}^1$, the probability of hitting E, starting from x, is given by

(2)
$$\Phi(x, E) = \int g(y - x)\mu(dy).$$

An obvious consequence of the regularity of these processes is the fact that $\Phi(x, E) = 1$ for all $x \in E$. This can be used to obtain information about the capacitary distribution μ on a large interval $[0, a]$. Several such results are given in [**4**] including the one mentioned at the end of the last section. We will need two of these here:

(3)
$$\mu[0, a] = b^{-1} \log a + O(1), \qquad \mu[0, a/2] = O(1).$$

Note that this means that most of the mass of the distribution is concentrated near the right end of the interval. A special case of (2) that will be useful is $\Phi(x, \{y\}) = Kg(y - x)$. Letting $y = x$ and using the regularity shows that $K = 1/g(0)$. The scaling property can be used to show that $g(0) = b - c$ so we have

(4)
$$\Phi(x, \{y\}) = g(y - x)/(b - c).$$

The first three lemmas will ultimately be used to provide information about where certain intervals are hit. For an arbitrary set E, T_E will be used to denote the first hitting time of E, i.e.,

$$T_E = \inf\{t > 0: X_t \in E\}$$

with the usual provision that $T_E = \infty$ if $X_t \notin E$ for all t. When we write the event $\{T_E < T_F\}$ it is to be understood that this means $\{T_E < T_F, T_E < \infty\}$ since $T_F \leqq \infty$ in any case.

LEMMA 1. *Let* $E = [a, a + \alpha]$, $F = [a + 2\alpha, \infty)$ *where* $\alpha = a/\log^\eta a$ *and* $0 < \eta < \infty$. *Then*

$$P\{T_F < T_E\} = O\Big(\frac{\log\log a}{\log a}\Big).$$

PROOF. First we have

(5)
$$\begin{aligned} P\{T_E < \infty\} &= P\{T_E < T_F\} + P\{T_F < T_E < \infty\} \\ &= P\{T_E < T_F\} + E\{I\{T_F < T_E\}P^{X_{T_F}}\{T_E < \infty\}\}. \end{aligned}$$

Now, for $x \in F$, since g is known to be ultimately monotone,

$$P^x\{T_E < \infty\} = \int_E g(y - x)\mu(dy) \leqq g(-\alpha)\mu(E) \to cb^{-1}$$

by (1) and (3). Therefore

$$\begin{aligned} P\{T_E < \infty\} &\leqq P\{T_E < T_F\} + (cb^{-1} + \varepsilon)\, P\{T_F < T_E\} \\ &= 1 - (1 - cb^{-1} - \varepsilon)\, P\{T_F < T_E\} \end{aligned}$$

since $\{T_E = T_F\} = \{T_E = T_F = \infty\} \subset \{T_F = \infty\}$ and this last event has probability zero. Since $c < b$, we must have $P\{T_F < T_E\} = O(1 - P\{T_E < \infty\})$. Thus it remains to be shown that $1 - P\{T_E < \infty\}$ is of the right order. This is done in similar situations in **[4]** but the proof in the present case is fairly easy so we will give it here. Partition the interval $[a, a + \alpha/2]$ into $\log \alpha$ intervals of length $\alpha/2 \log \alpha$. By (3), at least one of the small intervals must have capacitary measure $O(1/\log \alpha)$. Let x be the midpoint and u, v the endpoints of such an interval. Then

$$1 - P\{T_E < \infty\} = \int_E \{g(y - x) - g(y)\}\mu(dy) = \int_a^u + \int_u^v + \int_v^{a+\alpha/2} + \int_{a+\alpha/2}^{a+\alpha}$$

and we can now estimate each of these four integrals. The second integral is $O(1/\log \alpha)$ since the integrand is bounded and we chose the interval so the measure would be of this order. For the first and third integrals, the integrand is $O(1/\log \alpha)$ and measure is $O(1)$. The principal contribution comes from the final integral where the integrand is

$$\frac{b}{\log \alpha} + O\Big(\frac{\log\log \alpha}{\log^2 \alpha}\Big) - \frac{b}{\log a} + O\Big(\frac{1}{\log^2 a}\Big) = O\Big(\frac{\log\log a}{\log^2 a}\Big)$$

while the measure is $O(\log a)$. Thus we have

$$1 - P\{T_E < \infty\} = O\Big(\frac{\log\log a}{\log a}\Big).$$

LEMMA 2. *Let* $E = [-a, -\alpha]$, $F = [a, \infty)$ *where* $\alpha = a/\log^\eta a$ *and* $0 < \eta < \infty$. *Then*

$$P\{T_E < T_F\} = O\Big(\frac{\log\log a}{\log a}\Big).$$

PROOF. For $a \leqq x \leqq 2a$, we have

$$P^x\{T_E < \infty\} = \int_E g(y - x)\mu(dy) \geqq g(-3a)\mu(E) = cb^{-1} + O\Big(\frac{1}{\log a}\Big)$$

by (1) and (3). Using this in (5),

$$P\{T_E < \infty\} \geqq P\{T_E < T_F\} + P\{T_F < T_E, X_{T_F} \leqq 2a\}\left(\frac{c}{b} + O\left(\frac{1}{\log a}\right)\right).$$

But

$$P\{T_F < T_E, X_{T_F} > 2a\} \leqq P\{X_{T_F} > 2a\}$$
$$\leqq P\{T_{[a+2\alpha,\infty)} < T_{[a,a+\alpha]}\} = O\left(\frac{\log\log a}{\log a}\right)$$

by Lemma 1. Therefore

$$P\{T_F < T_E, X_{T_F} \leqq 2a\} = P\{T_F < T_E\} + O\left(\frac{\log\log a}{\log a}\right)$$
$$= 1 - P\{T_E < T_F\} + O\left(\frac{\log\log a}{\log a}\right),$$

the second equality following as in the last lemma since $T_F < \infty$ almost surely. Hence

$$P\{T_E < \infty\} \geqq P\{T_E < T_F\}\left(1 - \frac{c}{b}\right) + \frac{c}{b} + O\left(\frac{\log\log a}{\log a}\right).$$

Finally,

$$P\{T_E < \infty\} = \int_E g(y)\mu(dy) \leqq g(-\alpha)\mu(E) = \frac{c}{b} + O\left(\frac{\log\log a}{\log a}\right)$$

which is sufficient to complete the proof.

The next lemma shows that the hitting place of a large interval some distance to the left of the origin will usually be near the left end if it is hit at all. The reason is that the process will ordinarily jump well past the interval and then hit it during its steady movement back to the right.

LEMMA 3. *Let* $E = [-a, -a + \alpha]$, $F = [-a + 2\alpha, -\alpha]$ *where* $\alpha = a/\log^\eta a$ *and* $0 < \eta < \infty$. *Then*

$$P^z\{T_F < T_E\} = O\left(\frac{\log\log a}{\log a}\right)$$

uniformly in z *for* $z \geqq 0$.

PROOF. For $x \in F, y \in E$, $g(y - x) = c/\log a + O(\log\log a/\log^2 a)$ so that

$$P^x\{T_E < \infty\} = \int_E g(y - x)\mu(dy) = \frac{c}{b} + O\left(\frac{\log\log a}{\log a}\right).$$

Substituting this in (5) with the measure P replaced by P^z,

$$(6) \qquad P^z\{T_E < \infty\} = P^z\{T_E < T_F\} + \frac{c}{b} P^z\{T_F < T_E\} + O\left(\frac{\log\log a}{\log a}\right).$$

Interchanging the roles of E and F, the only difference is that, for $x \in E$, $P^x\{T_F < \infty\} = 1 + O(\log\log a/\log a)$ and so

$$P^z\{T_F < \infty\} = P^z\{T_F < T_E\} + P^z\{T_E < T_F\} + O\left(\frac{\log\log a}{\log a}\right).$$

Subtracting (6) from this, we have

$$\left(1 - \frac{c}{b}\right)P^z\{T_F < T_E\} = P^z\{T_F < \infty\} - P^z\{T_E < \infty\} + O\left(\frac{\log\log a}{\log a}\right).$$

Next, we consider the estimates

$$P^z\{T_F < \infty\} = \int_F g(y - z)\mu_F(dy) \leqq g(-\alpha - z)\mu_F(F),$$

$$P^z\{T_E < \infty\} = \int_E g(y - z)\mu_E(dy) \geqq g(-a - z)\mu_E(E).$$

Thus it will be sufficient to obtain upper bounds of order $\log\log a/\log a$ for $\{g(-\alpha - z) - g(-a - z)\}\mu_F(F)$ and $g(-a - z)\{\mu_F(F) - \mu_E(E)\}$. But $\mu_F(F) = O(\log a)$ and

$$\begin{aligned} g(-\alpha - z) - g(-a - z) &= \frac{c}{\log(\alpha + z)} - \frac{c}{\log(a + z)} + O\left(\frac{1}{\log^2 a}\right) \\ &= \frac{c\log(a + z)/(\alpha + z)}{\log(\alpha + z)\log(a + z)} + O\left(\frac{1}{\log^2 a}\right) \\ &\leqq \frac{c\log a/\alpha}{\log\alpha\log a} + O\left(\frac{1}{\log^2 a}\right) = O\left(\frac{\log\log a}{\log^2 a}\right). \end{aligned}$$

The other term is easy since $g(-a - z) = O(1/\log a)$ and

$$\mu_F(F) - \mu_E(E) = b^{-1}(\log a - \log\alpha) + O(1).$$

Next we need to introduce some more notation. Let $T_1 = 0$ and for $i > 1$ define T_i inductively by $T_i = T_{(B_{i-1},\infty)}$ where $A_i = X_{T_i}$, $B_i = A_i + a\log^{-\eta}a$, and η is a fixed real number in (0, 1). The A's, B's, and T's are random variables depending on a but this dependence will be suppressed in order to simplify the notation. We will use $R[t_1, t_2]$ to denote the range of the process on the restricted time interval $[t_1, t_2]$, i.e., $R[t_1, t_2] = \{x: X_t = x \text{ for some } t \in [t_1, t_2]\}$. Then let

$$Y_i = \left|R[0, T_{i+1}] \cap [A_i, B_i]\right| = \left|R[T_i, T_{i+1}] \cap [A_i, B_i]\right|.$$

By the strong Markov property, the Y_i are independent and identically distributed. Now we need to obtain some information about the distribution of the Y's.

LEMMA 4. *As $a \to \infty$, we have*

$$EY_1 \sim \frac{a}{\log^{1+\eta}a}, \qquad EY_1^2 = O\left(\frac{a^2}{\log^{2+2\eta}a}\right).$$

PROOF. Let $\alpha = a/\log^\eta a$. Note that $Y_1 = \left|R[0, T_2] \cap [0, \alpha]\right|$ and

(7) $$Y_1 + \left|R[T_2, \infty) \cap [0, \alpha]\right| = \left|R \cap [0, \alpha]\right| + \left|R[0, T_2] \cap R[T_2, \infty) \cap [0, \alpha]\right|.$$

Now $E\left|R[T_2, \infty) \cap [0, \alpha]\right| = EE^{X_{T_2}}\left|R \cap [0, \alpha]\right|$, and for $\alpha \leqq x \leqq 2\alpha$ we have, by (4),

$$E^x|R \cap [0, \alpha]| = E^x \int_0^\alpha I\{y \in R\}\, dy = \int_0^\alpha P^x\{y \in R\}\, dy$$

$$= \int_0^\alpha \frac{g(y-x)}{b-c}\, dy \sim \frac{c}{b-c} \frac{\alpha}{\log \alpha}.$$

For $x > 2\alpha$, we still have

$$E^x|R \cap [0, \alpha]| = \int_0^\alpha \frac{g(y-x)}{b-c}\, dy = O\left(\frac{\alpha}{\log \alpha}\right). \tag{8}$$

By Lemma 1, $P\{X_{T_2} \geqq 2\alpha\} \leqq P\{X_{T_2} \geqq \alpha + 2\alpha/\log \alpha\} = O(\log\log \alpha/\log \alpha)$ and so

$$E|R[T_2, \infty) \cap [0, \alpha]| \sim \frac{c}{b-c} \frac{\alpha}{\log \alpha}. \tag{9}$$

Of course,

$$E|R \cap [0, \alpha]| = E\int_0^\alpha I\{y \in R\}\, dy = \int_0^\alpha \frac{g(y)}{b-c}\, dy \sim \frac{b}{b-c} \frac{\alpha}{\log \alpha}. \tag{10}$$

For the final term in (7),

$$E|R[0, T_2] \cap R[T_2, \infty) \cap [0, \alpha]| = E\int_0^\alpha I\{y \in R[0, T_2], y \in R[T_2, \infty)\}\, dy$$

$$\leqq \beta + \int_0^{\alpha-\beta} P\{y \in R[0, T_2], y \in R[T_2, \infty)\}\, dy.$$

We will use $\beta = \alpha/\log^2\alpha$. For $0 \leqq y \leqq \alpha - \beta$,

$$P\{y \in R[0, T_2], y \in R[T_2, \infty)\} = E\{I\{y \in R[0, T_2]\}P^{X_{T_2}}\{y \in R\}\}$$

$$= O\left(\frac{1}{\log \alpha} P\{y \in R[0, T_2]\}\right) = O\left(\frac{g(y)}{\log \alpha}\right),$$

where the estimate used for $P^{X_{T_2}}\{y \in R\}$ is from $P^x\{y \in R\} = O(g(y-x))$ together with the fact that $X_{T_2} \geqq \alpha$. Thus we see that

$$E|R[0, T_2] \cap R[T_2, \infty) \cap [0, \alpha]| = O(\alpha/\log^2\alpha). \tag{11}$$

Using (9), (10), and (11) in (7) gives the asymptotic behavior of EY_1. For the estimate on the second moment of Y_1, write

$$EY_1^2 \leqq E|R \cap [0, \alpha]|^2 = E\int_0^\alpha \int_0^\alpha I\{x \in R, y \in R\}\, dx\, dy$$

$$= \int_0^\alpha \int_0^\alpha P\{x \in R, y \in R\}\, dx\, dy.$$

Then $P\{x \in R, y \in R\} = O(g(x)g(y-x) + g(y)g(x-y))$ and the result now follows easily by integration.

Let $N = \max\{i: B_i \leqq a\}$. Then $N \log^{-\eta} a \xrightarrow{P} 1$ but we can actually prove much more:

LEMMA 5. *As $a \to \infty$,*

$$P\{N \neq [\log^\eta a] \text{ or } [\log^\eta a] - 1\} = O(\log\log a/\log^{1-\eta} a).$$

PROOF. Since $B_i \geqq ia \log^{-\eta} a$ and $B_N \leqq a$, the inequality $N \leqq [\log^\eta a]$ is clear. Thus we need a lower bound. Again we let $\alpha = a/\log^\eta a$. By Lemma 1,

$$P\left\{A_i - B_{i-1} > \frac{2\alpha}{\log \alpha}\right\} = E\left\{P^{A_{i-1}}\left\{X_{T_i} - A_{i-1} > \alpha + \frac{2\alpha}{\log \alpha}\right\}\right\}$$

$$= P\left\{X_{T_2} > \alpha + \frac{2\alpha}{\log \alpha}\right\} = O\left(\frac{\log\log \alpha}{\log \alpha}\right).$$

Therefore

$$P\left\{A_i - B_{i-1} \leqq \frac{2\alpha}{\log \alpha} \text{ for all } i \leqq \log^\eta a\right\} = 1 - O\left(\frac{\log\log a}{\log^{1-\eta} a}\right)$$

and if this event occurs then $B_i - B_{i-1} \leqq \alpha + 2\alpha/\log \alpha$ for all $i \leqq \log^\eta a$, or $B_i \leqq i\alpha(1 + 2/\log \alpha)$ for all $i \leqq \log^\eta a$. It is now easy to check that this implies that $B_{[\log^\eta a]-1} \leqq a$ for a sufficiently large which is enough to complete the proof.

We now have the necessary preliminaries out of the way and are ready to give the proof of the theorem. The idea is that as X_t passes through the interval $[0, a]$, the measure of its range is approximately $a/\log a$. This will be shown first in Theorem 3. Once the process has left the interval $[0, a]$, the probability that it returns in any significant way is ρ. Furthermore, if it does return the point of return is near the origin. These facts lead to the geometric distribution and will be proved following the proof of Theorem 3.

Let $\tau = T_{(a,\infty)}$. Then we have

THEOREM 3. *As* $a \to \infty$

$$|R[0, \tau] \cap [0, a]| \log a/a \xrightarrow{P} 1.$$

REMARK. The argument given above to show that there cannot be almost sure convergence in the completely asymmetric case will also work to show that there cannot be almost sure convergence in Theorem 3 for any asymmetric Cauchy process.

PROOF. First we have that

$$|R[0, \tau] \cap [0, a]| \geqq Y_1 + Y_2 + \cdots + Y_N$$

and this is at least $Y_1 + Y_2 + \cdots + Y_{[\log^\eta a]-1}$ with probability approaching one. We can also check that the Y_i's satisfy the weak law of large numbers even though the distribution depends on a, for by Lemma 4,

$$P\{|Y_1 + \cdots + Y_k - kEY_1| \geqq \varepsilon kEY_1\} \leqq \frac{k \operatorname{Var} Y_1}{\varepsilon^2 k^2 (EY_1)^2} = O\left(\frac{1}{k}\right).$$

Thus we have with probability approaching one,

$$|R[0, \tau] \cap [0, a]| \geqq Y_1 + \cdots + Y_{[\log^\eta a]-1} \geqq (1 - \varepsilon) \log^\eta a \frac{a}{\log^{1+\eta} a} = (1 - \varepsilon)\frac{a}{\log a}.$$

To get the upper bound, we use the inequality

$$|R[0, \tau] \cap [0, a]| \leqq Y_1 + Y_2 + \cdots + Y_N + Y_{N+1} + Z_1 + Z_2 + \cdots + Z_N,$$

where $Z_i = |R[T_{i+1}, \tau] \cap [0, A_{i+1}]|$. The first part of the argument also shows that

$Y_1 + \cdots + Y_{N+1} \leqq (1 + \varepsilon)\, a/\log a$ with high probability so it will suffice to show that the remaining terms approach zero in probability when divided by $a/\log a$. We will do this by showing the normalized sum of the Z's goes to zero in L^1. Note that

$$EZ_i = E\{I\{T_{i+1} < \tau\}E^{A_{i+1}}\,|R[0, \tau] \cap [0, A_{i+1}]|\}.$$

Since $A_{i+1} \leqq a$ on $\{T_{i+1} < \tau\}$ we need to estimate $E^x|R[0, \tau] \cap [0, x]|$ for $0 \leqq x \leqq a$. Now, for any $\beta \in [0, x]$,

$$E^x|R[0, \tau] \cap [0,x]| \leqq \beta + E^x|R[0, \tau] \cap [0, x - \beta]| \tag{12}$$

and if $T = T_{[0,x-\beta]}$ we have

$$E^x|R[0, \tau] \cap [0, x - \beta]| = E^x\{I\{T < \tau\}E^{X_T}|R[0, \tau] \cap [0, x - \beta]|\}.$$

But, for any y,

$$E^y|R[0, \tau] \cap [0, x - \beta]| \leqq E^y|R \cap [0, x - \beta]| = O(a/\log a)$$

as in (8) and

$$P^x\{T < \tau\} \leqq P^x\{T_{[2x-2a,\, x-\beta]} < T_{[2a,\,\infty)}\}.$$

Lemma 2 now applies if we let $\beta = (2a - x)/\log^2(2a - x)$ to show that this probability is $O(\log\log a/\log a)$. Substituting these estimates back in (12), we have, for $0 \leqq x \leqq a$,

$$E^x|R[0, \tau] \cap [0, x]| = O(a \log\log a/\log^2 a)$$

and so $EZ_i = O(a \log\log a/\log^2 a)$ as well. It follows that

$$E(Z_1 + \cdots + Z_N)\frac{\log a}{a} = O\left(\frac{a \log\log a}{\log^2 a}\log^\eta a\,\frac{\log a}{a}\right) = O\left(\frac{\log\log a}{\log^{1-\eta} a}\right).$$

Now we are in a position to complete the proof of Theorem 1. First we define inductively two sequences of stopping times. Let $\sigma_1 = 0$ and then

$$\tau_i = \inf\{t \geqq \sigma_i\colon X_t > a\}, \qquad i \geqq 1,$$

$$\sigma_i = \inf\{t \geqq \tau_{i-1}\colon 0 \leqq X_t \leqq a - a/\log^2 a\}, \qquad i \geqq 2,$$

where we make the usual convention that any σ_i (and then all subsequent σ's and τ's) is infinite if there are no t's satisfying the conditions. Note that τ_1 is the same time that we called τ above. Let $M = \max\{i\colon \sigma_i < \infty\}$. On $\{M \geqq i\}$ we can define, again suppressing the dependence on a,

$$U_i = |R[\sigma_i, \tau_i] \cap [0, a]| \log a/a.$$

The statement of Theorem 3 is that $U_1 \xrightarrow{P} 1$. With a little work we shall see that the same applies to the other U's once they are defined. First, we note that since

$$|R[0, \tau] \cap [0, a]|\,\frac{\log a}{a} = |R[0, \tau] \cap [x, a]|\,\frac{\log a}{a} + O\left(\frac{x \log a}{a}\right),$$

it follows from Theorem 3 that, for $0 \leqq x \leqq 2a/\log^2 a$,

$$P^x\{|\,|R[0, \tau] \cap [0, a]| \log a a^{-1} - 1| > \varepsilon\} \to 0.$$

Letting $\alpha = a/\log^2 a$, we have

$$\begin{aligned} P\{|U_i - 1| > \varepsilon, M \geqq i\} &= E\Big\{I\{M \geqq i\}P^{X_{\sigma_i}}\Big\{\Big||R[0, \tau] \cap [0, a]|\frac{\log a}{a} - 1\Big| > \varepsilon\Big\}\Big\} \\ &\leqq P\{M \geqq i, X_{\sigma_i} > 2\alpha\} + o(1) \\ &\leqq E\{I\{M \geqq i - 1\}P^{X_{\tau_{i-1}}}\{T_{[2\alpha, a-\alpha]} < T_{[0, \alpha]}\}\} + o(1) \\ &= O(\log\log a/\log a) + o(1), \end{aligned}$$

the last bound being a consequence of Lemma 3. We must exercise a little care here since $\{M \geqq i\}$ is also changing with a. But we can state the last fact as $(U_i - 1)I\{M \geqq i\} \xrightarrow{P} 0$. Then

$$|(U_1 + \cdots + U_k - k)I\{M = k\}| \leqq \sum_{j=1}^{k} |U_j - 1| I\{M \geqq j\} \xrightarrow{P} 0$$

for each fixed $k \geqq 1$. From this it follows immediately that

$$U_1 + U_2 + \cdots + U_M - M \xrightarrow{P} 0, \tag{13}$$

provided that we show that $P\{M \geqq k\} \to 0$ uniformly in a. To see this we need to observe that

$$\Phi(x, [0, a - \alpha]) = \int_0^{a-\alpha} g(y - x)\mu(dy) \to \frac{c}{b} = \rho \quad \text{for } a \leqq x \leqq 3a, \tag{14}$$

and that by Lemma 1

$$\begin{aligned} P^z\{X_\tau \geqq 3a\} \leqq P^z\{T_{[a+z+2\alpha, \infty)} < T_{[a+z, a+z+\alpha]}\} = O\Big(\frac{\log\log a}{\log a}\Big) \\ \text{for} \quad 0 \leqq z \leqq a - \alpha. \end{aligned} \tag{15}$$

Then we have

$$P\{M \geqq i + 1\} = E\{I\{M \geqq i\}P^{X_{\sigma_i}}\{\sigma_{i+1} < \infty\}\}$$

and

$$P^{X_{\sigma_i}}\{\sigma_{i+1} < \infty\} = P^{X_{\sigma_i}}\{X_{\tau_i} < 3a, \sigma_{i+1} < \infty\} + P^{X_{\sigma_i}}\{X_{\tau_i} \geqq 3a, \sigma_{i+1} < \infty\}.$$

The second term goes to zero by (15) while the first can be written

$$E^{X_{\sigma_i}}\{I\{X_{\tau_i} < 3a\}P^{X_{\tau_i}}\{X_t \text{ hits } [0, a - \alpha]\}\}$$

and this tends to ρ by (14) and (15). Thus we have $P\{M \geqq i + 1\} \sim \rho P\{M \geqq i\}$ and since all the estimates are uniform in i,

$$P\{M \geqq i + 1\} \sim \rho^i P\{M \geqq 1\} = \rho^i.$$

Therefore, for a sufficiently large,

$$P\{M \geqq i + 1\} \leqq (\rho + \varepsilon)^i \tag{16}$$

which shows that $P\{M \geqq k\} \to 0$. At the same time we have seen that M has a limit distribution:

(17) $$P\{M = i\} = P\{M \geqq i\} - P\{M \geqq i + 1\} \to (1 - \rho)\rho^{i-1}.$$

To complete the proof we need to show that $U_1 + U_2 + \cdots + U_M$ is close to $|R \cap [0, a]| \log a/a$. To see this, first note that

$$R \cap [0, a] = \bigcup_{i=1}^{M} R[\sigma_i, \tau_i] \cap [0, a] \cup \bigcup_{i=1}^{M} R[\tau_i, \sigma_{i+1}) \cap [0, a].$$

The second union is included in the interval $[a - \alpha, a]$ and so its measure is no larger than α which is $o(a/\log a)$. Thus we need to compare

$$\left| \bigcup_{i=1}^{M} R[\sigma_i, \tau_i] \cap [0, a] \right| \frac{\log a}{a} \quad \text{and} \quad \sum_{i=1}^{M} |R[\sigma_i, \tau_i] \cap [0, a]| \frac{\log a}{a},$$

the latter sum being the sum of the U's. The difference is bounded by

(18) $$\sum_{i=1}^{M-1} |R[\sigma_i, \tau_i] \cap R[\sigma_{i+1}, \infty) \cap [0, a]| \frac{\log a}{a}$$

and it will suffice to show this goes to zero in probability or in L^1. Now

$$|R[\sigma_i, \tau_i] \cap R[\sigma_{i+1}, \infty) \cap [0, a]| = \int_0^{a-\alpha} I\{y \in R[\sigma_i, \tau_i], y \in R[\sigma_{i+1}, \infty)\} \, dy + O(\alpha)$$

and so we need to estimate for $0 \leqq y \leqq a - \alpha$

$$\begin{aligned} &E\{I\{M \geqq i + 1\} I\{y \in R[\sigma_i, \tau_i], y \in R[\sigma_{i+1}, \infty)\}\} \\ &\leqq E\{I\{M \geqq i\} I\{y \in R[\sigma_i, \tau_i]\} P^{X_{\tau_i}}\{y \in R\}\} \\ &= O(1/\log \alpha) P\{M \geqq i, y \in R[\sigma_i, \tau_i]\} \\ &= O(1/\log \alpha) E\{I\{M \geqq i\} P^{X_{\sigma_i}}\{y \in R\}\}. \end{aligned}$$

For any $x \in [0, a - \alpha]$, we have

$$\int_0^{a-\alpha} P^x\{y \in R\} \, dy = O\left(\int_0^{a-\alpha} g(y - x) \, dy\right) = O\left(\frac{a}{\log a}\right).$$

Therefore

$$E\{I\{M \geqq i + 1\} |R[\sigma_i, \tau_i] \cap R[\sigma_{i+1}, \infty) \cap [0, a]|\} = O((a/\log^2 a) P\{M \geqq i\}).$$

Summing these estimates using (16), we see that the expectation of (18) is $O(1/\log a)$. To summarize, putting this last part together with (13) shows that

$$|R \cap [0, a]| \log a a^{-1} - M \xrightarrow{P} 0,$$

and the limit distribution of M has been obtained in (17).

The interested reader should now be able to prove the second statement of Theorem 1.

References

1. R. K. Getoor and H. Kesten, *Continuity of local times for Markov processes*, Compositio Math. **24** (1972), 277–303. MR **46** #10075.

2. P. W. Millar and L. T. Tran, *Unbounded local times*, Z. Wahrscheinlichkeitstheorie und Verw. Gebiete **30** (1974), 87–92. MR **52** #1909.

3. S. C. Port and C. J. Stone, *The asymmetric Cauchy processes on the line*, Ann. Math. Statist. **40** (1969), 137–143. MR **38** #3922.
4. W. E. Pruitt and S. J. Taylor, *Asymmetric Cauchy processes* (to appear).

UNIVERSITY OF MINNESOTA

LIVERPOOL UNIVERSITY

Proceedings of Symposia in Pure Mathematics
Volume 31, 1977

THE MARTIN BOUNDARY OF A RECURRENT RANDOM WALK HAS ONE OR TWO POINTS

D. REVUZ

This is a report on a joint work with A. Brunel. In the forthcoming paper [4] we describe the Martin boundary for all recurrent random walks on countable groups, thus answering in the affirmative a question of Kesten [5]. In the following I shall present this result and try to give a rough outline of its proof. I want to thank A. Brunel for allowing me to use our joint results herein.

I. Notation and statement of the problem. Throughout the sequel we consider a countable group G with neutral element e and a probability measure μ on G such that

(i) G is generated by the support of μ;

(ii) $\sum_0^\infty \mu^n(e) = +\infty$, where μ^n stands for the nth convolution power of μ. We study the left random walk of law μ on G, that is the Markov chain M with transition probability $P(x,\cdot) = \mu * \varepsilon_x$, $x \in G$. This is clearly an irreducible recurrent chain and it is then well known that it has no potential operator in the classical sense. However it was shown by Kesten [5] that the following limit exists:

$$a(x) = \lim_{n\to\infty} \sum_0^n [P_k(e, e) - P_k(x, e)] = \lim_{n\to\infty} \sum_0^n [\mu^k(e) - \mu^k(x^{-1})]$$

where x^{-1} is the inverse of x in G. The existence of this limit permits us to define a substitute to the potential kernel. Indeed, the function a enjoys the following properties: (i) $a \geqq 0$, $a(e) = 0$; (ii) $Pa = a + 1_e$, and if f is a special function in the sense of Neveu [7], [8], in particular if f is a function with finite support, then the function Af defined by

$$Af(x) = \sum_{y\in G} a(xy^{-1}) f(y)$$

AMS (MOS) subject classifications (1970). Primary 60J15.

is finite and is a solution of the Poisson equation $P\varphi = \varphi - f$.

Another kernel which works more or less as a potential kernel for M is the chain obtained by killing M when it first hits e. We have

$$G^e = \sum_0^\infty Q^n \quad \text{where } Q = 1_{\{e\}^c} P 1_{\{e\}^c}$$

and if f is special then $PG^e f = G^e f + m(f)1_e - f$ where m is the counting measure of G, and $G^e Pf = G^e f + f(e) - f$.

The two kernels G^e and A are related by the identity $G^e(x, y) = a(x) + a(y^{-1}) - a(xy^{-1})$ which incidentally proves that a is a subadditive function on G, a fact of paramount importance in many proofs. All these results are shown or commented upon in **[5]**, **[10]** and **[11]**.

We steadfastly have also to deal with the dual random walk $\hat{M}$ and as usual every item pertaining to $\hat{M}$ will be designated by the hat "^" or by the prefix co. For instance

$$\hat{a}(x) = \lim_n \sum_0^n [\hat{P}_k(e, e) - \hat{P}_k(x, e)] = \lim_n \sum_0^n [P_k(e, e) - P_k(x^{-1}, e)] = a(x^{-1}),$$

and the functions which are special for $\hat{M}$ will be referred to as cospecial.

We are now ready to state the problem we are aiming to solve. The recurrent Martin boundary serves several purposes as is shown in the work by Orey and by Kemeny and Snell (see **[10**, Chapter 8]). In particular it leads to an integral representation of the solutions of the Poisson equation. It is defined as the (transient) Martin boundary of the chain obtained by killing M when it first hits e. It is therefore in one-to-one correspondence with the set of limit functions of $G^e(x, \cdot)$ as x tends to Δ, the point at infinity in G. One may also view this as the recurrent counterpart of renewal theory since the set of limit functions of $G^e(x, \cdot)$ is in one-to-one correspondence with the set of limit functions of $\hat{a}(\cdot) - \hat{a}(\cdot x^{-1})$ which is the copotential of $1_e - 1_x$.

To summarize, we shall be interested in the following problem: *Find the set of limit functions of $G^e(x, \cdot)$ as x tends to Δ.*

This problem was solved by Kesten and Spitzer for abelian groups and tackled by Kesten for nonabelian groups. Kesten proved the following results:

THEOREM 1 (KESTEN). *There are one, two or infinitely many limit functions.*

THEOREM 2 (KESTEN). *There are two limit functions if and only if there exists an infinite cyclic subgroup G_0 of finite index in G such that the random walk induced on G_0 has a finite variance.*

Finally he stated the following

CONJECTURE. *There can be only one or two limit functions.*

It is this conjecture that we claim to have solved in **[4]**. Furthermore we can also compute the limit functions.

II. Statement of the result. To state our results we must introduce a classification of random walks given in **[1]** which generalizes the classification of Spitzer **[11]**

for random walks on the integer lattices. The classification rests on the following result of [**1**].

PROPOSITION 1. *If φ is a function such that*
(i) $\varphi(e) = 0, \varphi \geqq 0$,
(ii) $P\varphi = \varphi + 1_e$,
(iii) $\varphi(\cdot) - \varphi(\cdot x)$ *is bounded for every x in G,*
then $\varphi = a + \chi$ where χ is a homomorphism of G into the real numbers such that $P|\chi| - |\chi|$ is m-integrable.

The existence of a nontrivial χ satisfying these conditions demands that the group G_0 and the law μ of M possess a special structure.

PROPOSITION 2. *If there exists a nontrivial homomorphism χ such that $P|\chi| - |\chi|$ is m-integrable, then the normal subgroup* Ker χ *is finite and $G/\text{Ker}\,\chi$ is a cyclic group.*

Let t be an element such that $G/\text{Ker}\,\chi$ is equal to $\{t^n, n \in \boldsymbol{Z}\}$. We define X to be the homomorphism of G into $\boldsymbol{Z}$ given by $X(t^n) = n$. The homomorphisms χ will be real multiples of X. Since the random walk is recurrent we have $\int_G X\,d\mu = 0$ and we set $\sigma^2 = \int_G X^2\,d\mu$. One can then prove

PROPOSITION 3. *There exists a nontrivial homomorphism χ such that $P|\chi| - |\chi|$ is m-integrable if and only if the group G has the aforementioned structure and if $\sigma^2 < +\infty$.*

DEFINITION 1. In the case just described the random walk M will be said of type II; otherwise it is said of type I.

For type I walks the function a is the only one to enjoy properties (i)—(iii) of Proposition 1.

We see that the case of type II random walks is settled by Kesten's Theorem 2 above. In fact, owing to the special structure of G, one can show

THEOREM 4. *If M is of type* II, *then for every special function f we have*

$$\lim_{x\to\pm\infty} G^e f(x) = \int \hat{a} f\,dm \pm \sigma^{-2}\int Xf\,dm$$

where $x \to \pm\infty$ means $X(x) \to \pm\infty$.

This result does not cover all the cases of Kesten's Theorem 2. We introduce therefore the following

DEFINITION 2. A random walk will be said of type I′ if it is type I and if there exists a normal subgroup G_0 of index 2 in G such that the random walk induced on G_0 is type II.

If M is type I′ we shall call X_0 the corresponding homomorphism on G_0 and set $X = P_{G_0} X_0$ where P_{G_0} is the usual balayage operator related to G_0. Using Theorem 4 one can show

THEOREM 5. *If M is type* I′, *then for every special function we have*

$$\lim_{x\to\pm\infty} G^e f(x) = \int \hat{a} f\,dm \pm \sigma^{-2}\int Xf\,dm$$

where $x \to \pm\infty$ means $X(x) \to \pm\infty$ and σ^2 is the variance of the random walk induced on G_0.

From now on type I will mean type I but not type I′. Our problem is finally settled by

THEOREM 6. *If M is type* I *then for every special and cospecial function f*

$$\lim_{x \to \Delta} G^e f(x) = \int \hat{a} f \, dm.$$

It was shown in [3] that f cannot be taken merely special if one wants to state the result in full generality.

Together with Theorems 4 and 5 the preceding result shows that Kesten's conjecture is valid. The first two theorems correspond to the case of two limit functions, the latter to the case of one limit function. We are now going to sketch the proof.

III. Outline of the proof. The case of type II and type I′ random walks is solved in [2] by direct methods which use the fact that the group G is then not too different from an abelian group. We will therefore concentrate below on type I random walks.

In the following we shall assume that a does not vanish outside $\{e\}$. The case where it does so is easily handled afterwards. We then define $\psi = a/\hat{a}$ outside $\{e\}$ and $\psi(e) = 1$. Plainly $\psi(x) = (\psi(x^{-1}))^{-1}$ and it may be shown that ψ is a bounded function if and only if the sets of special and cospecial function are equal. The reason to consider the function ψ is the following result which follows from the recurrent maximum principle.

PROPOSITION 1. *If $f \in \mathscr{L}^1(\hat{a} \cdot m)$, in particular if f is special, then*

$$G^e f \leqq (1 + \psi) \int \hat{a} f \, dm.$$

Let us now suppose that ψ is bounded from above by a constant L. Let $\mathscr{U}$ be any ultrafilter finer than the set of neighborhoods of Δ; for every $f \in \mathscr{L}^1_+(\hat{a} \cdot m)$ we have

$$\lim_{\mathscr{U}} G^e f \leqq (1 + L) \int \hat{a} f \, dm.$$

There exists therefore a function $\varphi_{\mathscr{U}}$ bounded from above by a multiple of $\hat{a}$ such that $\lim_{\mathscr{U}} G^e f = \int f \cdot \varphi_{\mathscr{U}} \, dm$. The space $\mathscr{L}^1_+(\hat{a} \cdot m)$ is invariant by P and therefore we also have $\lim_{\mathscr{U}} G^e Pf = \int Pf \cdot \varphi_{\mathscr{U}} \, dm$. Since $G^e Pf = G^e f - f + f(e)$ we obtain

$$\int Pf \cdot \varphi_{\mathscr{U}} \, dm = \int f \varphi_{\mathscr{U}} \, dm + f(e)$$

which yields $\hat{P}\varphi_{\mathscr{U}} = \varphi_{\mathscr{U}} + 1_e$. Moreover $\varphi_{\mathscr{U}}(e) = 0$ obviously. Let us summarize the properties of $\varphi_{\mathscr{U}}$:

(∗)
(i) $\varphi_{\mathscr{U}} \geqq 0$ and $\varphi_{\mathscr{U}}(e) = 0$,
(ii) $\hat{P}\varphi_{\mathscr{U}} = \varphi_{\mathscr{U}} + 1_e$,
(iii) $\varphi_{\mathscr{U}}$ is majorized by a multiple of $\hat{a}$.

The function $\hat{a}$ satisfies the set (∗) of conditions. If perchance $\hat{a}$ were the only function to satisfy (∗) then we would get $\lim_{x\to\Delta} G^e f(x) = \int \hat{a} f\, dm$ for every $f \in \mathscr{L}^1_+(\hat{a}\cdot m)$ and we would be finished. Unfortunately, as the example of type I′ random walks shows, there may exist functions other than $\hat{a}$ satisfying (∗) and we must now try to characterize the situation where $\hat{a}$ is the only solution of (∗).

We therefore work with a random walk which is not of type II and study the cone C of functions f such that

$$f(e) = 0, \quad f \geqq 0, \quad Pf \leqq f + t(f)\, 1_e,$$

where $t(f)$ is a constant. The set $K = \{f\colon t(f) \leqq 1\}$ is convex, and compact and metrizable for the topology of pointwise convergence. The function a is in K and has therefore a unique Choquet representation by means of extremal functions of K. We prove the following result which is the key to the solution of Kesten's conjecture.

THEOREM 2. *Either the function a is extremal in K or the random walk is type* I′.

OUTLINE OF THE PROOF. Let $a = \int f_\alpha\, d\nu(\alpha)$ be the integral representation of a. The function f_α satisfies the equality $Pf_\alpha = f_\alpha + 1_e$. It may be shown, and this is a crucial point, that at least one of the functions f_α, say f_θ, is not mutually singular with a. In other words since f_θ is extremal in C, if $f_\theta \wedge a$ is the infimum of f_θ and a for the intrinsic order of the cone C, there exists a number k such that $f_\theta \wedge a = kf_\theta$; hence $f_\theta \leqq k^{-1}a$.

Now by defining

$$T_x f(y) = f(yx) - f(x) + t(f)\, G^e(y, x^{-1})$$

we turn G into a group of affine transformations of the set H of functions such that: (i) $f \geqq 0, f(e) = 0$, (ii) $Pf = f + t(f)\, 1_e$, into itself. Furthermore $t(T_x f) = t(f)$ and a is invariant by T_x.

For every x in G, we have $T_x f_\theta \leqq k^{-1}a$ and $T_x f_\theta$ is an extremal element of H. Since H is a complete lattice, the set $\{T_x f_\theta, x \in G\}$ has a lower upper bound Φ in H. It is easily seen that $\Phi(\cdot) - \Phi(\cdot x)$ is bounded for every x in G and by the results in §II it follows that Φ is a multiple of a. On the other hand, the $T_x f_\theta$, $x \in G$, are also extremal; hence Φ is equal to their sum. Since Φ is finite we may derive that there exists a finite number of points in G, say $x_1, \cdots, x_n$ and a constant L such that

$$a = L(f_\theta + T_{x_1} f_\theta + T_{x_2} f_\theta + \cdots + T_{x_n} f_\theta).$$

Since $t(a) = t(f_\theta) = 1$, it follows that $L = (n + 1)^{-1}$.

Furthermore it may be shown that for a function f in H with $t(f) \leqq 1$ we have $f \leqq a + \hat{a}$; on the other hand, if f and g are in H, their supremum $f \vee g$ for the intrinsic order is equal to $\sup(f, g) + G^e\Psi$, that is the sum of their supremum for the usual order and of a potential $G^e\Psi$ where Ψ is m-integrable. As a result $(n + 1)a \leqq a + \hat{a} + G^e\Psi$. But clearly $\overline{\lim}_{x\to\Delta} (a/(a + \hat{a})) \geqq \frac{1}{2}$ and it can be shown that the integrability of Ψ entails $\lim_{x\to\Delta} G^e\Psi/(a + \hat{a}) = 0$. Consequently, by passing to the limit in the above inequality after dividing by $a + \hat{a}$ we get $(n + 1)/2 \leqq 1$. This entails that only two cases are possible, $n = 0$ and $n = 1$. For $n = 0, f_\theta = a$, hence a is extremal; if $n = 1$ then the set $G_0 = \{x \in G\colon T_x f_\theta = f_\theta\}$ is a normal subgroup of index 2 in G and it is straightforward if tedious to show that the random walk is type I′.

We may now give an

OUTLINE OF THE PROOF OF THEOREM 6 OF §II. Plainly we have only to deal with random walks of type I which are not type I′, that is we may assume that $\hat{a}$ is the only function satisfying the set (∗) of conditions. If $\mathscr{V}$ is bounded, we then know by the reasoning of the beginning of the section that Theorem 6 is true. Our method of proof is then as follows.

(A) We prove directly Theorem 6 for the case where every element in G is of finite order. (We conjecture that in this case $\mathscr{V}$ is bounded but we have not been able to prove it.)

(B) If there exists an element of finite order we prove that either $\mathscr{V}$ is bounded or G has a special structure which permits us to prove Theorem 6 directly. For instance if G is finitely generated, has an element of infinite order and $\mathscr{V}$ is not bounded, then there exists a finite normal subgroup G' such that G/G' is abelian; thus G is not too different from an abelian group, and as we have already pointed out it is then possible to prove Theorem 6 directly.

REFERENCES

1. A. Brunel and D. Revuz, *Marches récurrentes au sens de Harris sur les groupes localement compacts*. I, Ann. Sci. École Norm. Sup. (4) **7** (1974), 273–310 (1975). MR **51** #1970.

2. ———, *Marches récurrentes au sens de Harris sur les groupes localement compacts*. II, Bull. Soc. Math. France **104** (1976), 3–31.

3. ———, *Marches récurrentes au sens de Harris sur les groupes localement compacts*. III (to appear).

4. ———, *Marches récurrentes au sens de Harris sur les groupes localement compacts*. IV (to appear).

5. H. Kesten, *The Martin boundary of recurrent random walks on countable groups*, Proc. Fifth Berkeley Sympos. Math. Statist. and Probability (Berkeley, Calif., 1965/66), Vol. II: Contributions to Probability Theory, Part 2, Univ. California Press, Berkeley, Calif., 1967, pp. 51–74. MR **35** #4988.

6. H. Kesten and F. Spitzer, *Random walk on countably infinite Abelian groups*, Acta. Math. **114** (1965), 237–265. MR **33** #3366.

7. J. Neveu, *Potentiels markoviens discrets*, Ann. Univ. de Clermont **24** (1964), 37–89.

8. ———, *Potentiel markovien récurrent des chaînes de Harris*, Ann. Inst. Fourier (Grenoble) (2) **22** (1972), 85–130.

9. S. Orey, *Potential kernels for recurrent Markov chains*, J. Math. Anal. Appl. **8** (1964), 104–132. MR **28** #3482.

10. D. Revuz, *Markov chains*, North-Holland, Amsterdam, 1975.

11. F. L. Spitzer, *Principles of random walk*, Van Nostrand, Princeton, N. J., 1964. MR **30** #1521.

UNIVERSITÉ PARIS VII

Proceedings of Symposia in Pure Mathematics
Volume 31, 1977

THE COFINE TOPOLOGY REVISITED

JOHN WALSH

Introduction. Suppose X and $\hat{X}$ are strong Markov processes in duality. Then both X and $\hat{X}$ have a fine topology. We will speak of the fine topology of $\hat{X}$ as the *cofine topology*. There is some interest in describing this topology in terms of X, not $\hat{X}$, because of the following question: "Given a strong Markov process X, when does there exist a strong Markov process $\hat{X}$ in duality with X?" The answer is that, modulo some technical considerations, this happens if and only if there exists a cofine topology [18]. Of course, to make sense of this, one must define the cofine topology without using $\hat{X}$. Such a description was given in [20]. Unfortunately, this description was unwieldy, to say the least, since it involved the entire collection of sample paths of X. We have felt for some time that one should be able to give a purely analytic construction of the cofine topology, involving only the excessive functions of the process. This article is an attempt at exactly that.

A strong hint as to where to look was provided by J. C. Taylor [19] in a study of Brelot sheaves. He showed that if a Green's function for a given Brelot sheaf exists, then one can construct a dual process. The fine topology for this dual process would then give us a cofine topology. This suggests that one should be able to construct a cofine topology directly from the Green's function.

The third ingredient of this study is Mokobodzki's theorem on the integral representation of excessive functions [13], [15], which makes it possible to give simple and general conditions which ensure the existence of a Green's function. We do this in Appendix A.

The principal tools we use are Doob's "h-path processes", which we call h-transforms here, and the close relation of these processes with co-optional times (or L-times), and in particular, with last exit times. Because these processes are less familiar than they deserve to be, we have included a review in §1 of those properties we shall need.

§2 revolves around the question of Green's potentials and the study of the func-

AMS (MOS) subject classifications (1970). Primary 60J40, 60J45.

tion $p_A(x) = P^x\{\exists\, t \ni X_{t-} \in A\}$. The results will seem both familiar and unfamiliar to the adepts of duality theory; the unfamiliarity results from the fact that familiar results appear with an "X_t" replaced by its left limit "X_{t-}". In the usual theory of two standard processes in duality, cf. [1], it turns out that the process X is quasi-left continuous and satisfies hypothesis (B) of Hunt. The effect of this is to make first hitting times of X_t and X_{t-} coincide (see Lemma 5.2), and our results then reduce to the usual ones.

The cofine topology is defined directly in terms of the functions $p_A(x)$ in §3, and we show that it satisfies hypothesis (CF) below. The last two sections are devoted to examples and to tracing the connection of the cofine topology with two other related topologies which have been proposed.

Before embarking on our quest after the elusive cofine topology, let us look at a sketch of its outlines, as given in [20]. Stripped of some unnecessary conditions, it is:

HYPOTHESIS (CF). There exists a topology $\mathscr{T}$ on E, finer than the initial topology, satisfying

(CF1) Let f be Borel and $\mathscr{T}$-continuous on E. Then $s \to f(X_{s-})$ is left continuous.

(CF2) If f is Borel and if $s \to f(X_{s-})$ has left limits P^x-a.s. on $(0, \infty)$ for all x, then $\bar{f}(X_{s-})$ is left continuous, where $\bar{f}(y) = \mathscr{T}\text{-}\lim\sup_{x \to y} f(x)$.

Moreover, $\bar{f}(X_{s-}) = \lim_{r \uparrow s} f(X_{r-})$ P^x-a.s., all $x \in E$.

As promised, this is unwieldy. To make sense out of it, one should remember that the dual process $\hat{X}$ is essentially the original process X reversed in time and made right continuous, and that we are describing its fine topology. Thus, for instance, if f is fine continuous for $\hat{X}$, $s \to f(\hat{X}_s)$ will be right continuous. Remembering the change of time sense, this means that $f(X_{s-})$ must be left continuous, which is just (CF1).

1. *h*-transforms. Let E be a locally compact separable metric space, to which we adjoin an isolated point δ to act as a cemetery. Let $X = (\Omega, \mathfrak{F}, \mathfrak{F}_t, X_t, \theta_t, P^x)$ be a strong Markov process on $E \cup \delta$ whose paths are right continuous and have left limits. We will take the space Ω above to be the space of all functions ω: $[0, \infty) \to E \cup \delta$ which are right continuous, admit a lifetime $\zeta(\omega)$, and which have left limits everywhere except possibly at ζ. We remark that the process X has a left limit at ζ, but some of the h-transforms considered below may not.

Let (P_t) be the semigroup of X. We suppose P_t is $\mathscr{B}$-measurable, where $\mathscr{B}$ is the topological Borel field of E.

If h is an excessive function, let $E_h = \{x \in E : 0 < h(x) < \infty\}$, and define the *h-transform semigroup* $({}_hP_t)$ by

$$
{}_hP_t(x, dy) = \begin{cases} h(x)^{-1}P_t(x, dy)h(y) & \text{if } x \in E_h, \\ 0 & \text{otherwise.} \end{cases} \tag{1.1}
$$

${}_hP_t$ is a sub-Markov semigroup on E; we extend it to a Markov semigroup on $E \cup \delta$ as usual.

Doob introduced these semigroups in [5] and [6], and established the properties of the associated Markov process X^h, which we will call the *h-transform* of X. He did this principally for Brownian motion, but his results and proofs extend to the general case with surprisingly little change. These h-transforms are well understood, but not well enough known to figure in the standard reference books. We make such heavy use of them in this article that we thought it worthwhile to summarize the standard results in this section. We refer to **[5]** and **[11]** for proofs. A word of warning: we follow the custom of supposing that all functions vanish at δ, unless it is explicitly stated otherwise.

A function f is h-excessive if it is excessive for ${}_hP_t$. h-excessive functions are essentially ratios of excessive functions, in the following sense.

PROPOSITION 1.1. *Let v be universally measurable. If v is h-excessive, then $v = 0$ on $E - E_h$ and there exists an excessive function u such that $u = 0$ on $\{h = 0\}$ and $u = hv$ on E_h. Conversely, if u is excessive and v satisfies*

(a) *$v = 0$ on $E - E_h$,*

(b) *$u = vh$ on E_h,*

then v is h-supermedian and ${}_hP_0 v$ is h-excessive.

Note. If there are no branching points, Proposition 1.1 simplifies: v is h-excessive iff there exists an excessive u satisfying (a) and (b).

The Markov process X^h corresponding to ${}_hP_t$ will have a right continuous version whose paths have left limits, except possibly at ζ. Thus X^h can be canonically defined on Ω, and we write

$$X^h = (\Omega, \mathfrak{F}, \mathfrak{F}_t, X_t, \theta_t, {}_hP^x).$$

We denote the natural (i.e., uncompleted) fields by $\mathfrak{F}^0$. The relation of X and X^h is given by the basic formula:

$$(1.2) \qquad {}_hP^x\{\Lambda; \zeta > t\} = E^x\{\Lambda; h(x_t)\}/h(x) \quad \text{if } x \in E_h \text{ and } \Lambda \in \mathfrak{F}^0_t.$$

Thus ${}_hP^x \ll P^x$ on $\mathscr{F}^0_t \cap \{\zeta > t\}$, and the R-N derivative is $h(X_t)$.

Let $\mathscr{F}^\mu$ be the completion of $\mathscr{F}^0$ relative to a measure μ on $\mathscr{F}^0$, and let $\mathscr{F}^\mu_t$ be $\mathscr{F}^0_t$ augmented by all sets of μ-measure zero of $\mathscr{F}^\mu$. Let $\mathscr{F}^* = \bigcap_\mu \mathscr{F}^\mu$ and $\mathscr{F}^*_t = \bigcap_\mu \mathscr{F}^\mu_t$, where the intersection is over all probability measures μ on $\mathscr{F}^0$.

THEOREM 1.2. *X^h is a strong Markov process. If T is a stopping time and $\Lambda \in \mathscr{F}^*_T$, then for $x \in E_h$*

$$(1.3) \qquad {}_hP^x\{\Lambda; \zeta > T\} = h(x)^{-1}E^x\{\Lambda; h(X_T)\}.$$

Let $\{u_\alpha, \alpha \in I\}$ be a family of excessive functions, such that $(x, \alpha) \to u_\alpha(x)$ has the necessary measurability properties, and define an excessive function h by

$$(1.4) \qquad h(x) = \int u_\alpha(x)\nu(d\alpha).$$

PROPOSITION 1.3. *Let ${}_\alpha P = {}_{u_\alpha}P$ and define h by* (1.4). *Then*

$$(1.5) \qquad {}_hP^x\{\Lambda\} = h(x)^{-1} \int u_\alpha(x)\,{}_\alpha P^x\{\Lambda\}\nu(d\alpha), \qquad \Lambda \in \mathscr{F}^*.$$

Let $A \subset E$ and define the *reduit operator* R_A as follows. If u is excessive, set

$$\underline{u}(x) = \inf\{v(x) \colon v \geqq u \text{ on } A, \ v \text{ excessive}\}.$$

If there exists a reference measure, then $\underline{u}$ will be supermedian, and we can define $R_A u$ to be its excessive regularization:

$$R_A u(x) = \lim_{t \downarrow 0} P_t \underline{u}(x).$$

Let $T_A = \inf\{t > 0: X_t \in A\}$. By Hunt's balayage theorem as extended by Shih [**17**], and Proposition 1.3:

PROPOSITION 1.4. *If $A \in \mathscr{B}$, u is excessive, and $u(x) < \infty$, then*

$$R_A u(x) = E^x\{u(X_{T_A})\} = u(x)_u P^x\{T_A < \infty\}. \tag{1.6}$$

An excessive function h is *minimal* if, whenever u and v are excessive and $h = u + v$, both u and v are proportional to h.

We define the *invariant field* $\mathscr{I}$ to be the set of $\Lambda \in \mathfrak{F}^*$ such that for each t,

$$\theta_t^{-1}(\Lambda \cap \{\zeta > 0\}) = \Lambda \cap \{\zeta > t\}.$$

The invariant field is connected with the behavior of X at its lifetime; events such as $\{\zeta = \infty\}$ and $\{X_{\zeta-} \in B\}$ are invariant. Heuristically, an h-transform of X can be thought of as the process X conditioned to have a given behavior at ζ. This behavior is particularly transparent when h is minimal.

PROPOSITION 1.5. *If h is excessive and $\Lambda \in \mathscr{I}$, the function $u(x) = {}_hP^x\{\Lambda; \zeta > 0\}$ is h-excessive. If h is minimal, u is constant on E_h, and $\mathscr{I}$ is trivial under ${}_hP^x$.*

Define the last exit M_A from $A \subset E$ by

$$M_A = \sup\{t: X_t \in A\},$$

where the supremum of the empty set is zero. If A is Borel, $\{M_A < \zeta\} \in \mathscr{I}$. If h is minimal excessive, we say A is *h-thin* if ${}_hP^x\{M_A < \zeta\} = 1$, all $x \in E_h$.

PROPOSITION 1.6. *If $A \in \mathscr{B}$ and h is minimal, then A is h-thin iff $R_A h \neq h$.*

We say a minimal excessive function h has a *pole* at $y \in E$ if, whenever V is a neighborhood of y, V^c is h-thin. Here is one reason for the importance of poles.

THEOREM 1.7. *Suppose h is minimal excessive. Then h has a pole at y iff ${}_hP^x\{X_{\zeta-} = y\} = 1, \forall\, x \in E_h$. If h does not have a pole at y, ${}_hP^x\{X_{\zeta-} = y\} \equiv 0$.*

There is a close connection between h-transforms and last exit times, or more generally, co-optional times, which we will be continually exploiting below. A random variable L, $0 \leqq L \leqq \infty$, is *co-optional* if, for each t, $L \circ \theta_t = (L - t) \vee 0$. If L is co-optional, define an excessive function C_L by

$$C_L(x) = P^x\{L > 0\}.$$

Define X^L, the *process killed at* L by

$$X_t^L = \begin{cases} X_t & \text{if } t < L, \\ \delta & \text{if } t \geqq L. \end{cases}$$

THEOREM 1.8 [**14**]. *X^L is a C_L-transform of X.*

2. Capacity. We assume as in §1 that X is a right continuous strong Markov process having left limits and with a Borel measurable semigroup. There are three additional hypotheses which will be in force for the remainder of the paper:

(T) X is transient, i.e., for each compact $K \subset E$, $P^x\{M_K < \infty\} = 1$;

(L) there exists a reference probability measure μ_0;

(G) there exists a left polar Borel set $N_1 \subset E$ such that if $y \in E - N_1$, there exists *at most* one minimal excessive potential (up to proportionality) having pole y.

Of these hypotheses, (T) is purely for convenience; if X were not transient, we could kill it at an exponential time and make it so without affecting what follows. From the analytic point of view, this would correspond to looking at the λ-excessive functions rather than the excessive functions. Hypothesis (L) is familiar. It is hypothesis (G) (for Green's function) which makes things work in this paper. It is a minor modification of R.-M. Hervé's hypothesis of proportionality.

We show in Appendix A that a *Green's function* exists under these hypotheses, i.e., a family $\{g_y: y \in E - N\}$, where N is Borel and contains N_1, and

(i) g_y is minimal excessive with pole y;

(ii) $(x, y) \to g_y(x)$ is $\mathscr{B} \times \mathscr{B}$ measurable.

We say that an excessive function h of the form $h(x) = \int g_y(x)\mu(dy)$ is a *Green's potential* and will often denote it by $G\mu$. The *support* of $G\mu$ is the support of μ. We will assume that the Green's function comes from the cap C_1, so that from Theorem A8 of the Appendix we get the following representation theorem.

PROPOSITION 2.0. *Let $A \in \mathscr{B}$ and let h be a bounded excessive function such that*

$$(2.1) \qquad {}_hP^x\{X_{\zeta-} \in A\} = 1, \quad \text{all } x \in E_h.$$

Then h is the Green's potential of a measure μ, and $\mu(E - A) = 0$.

A converse of this theorem is the following.

PROPOSITION 2.1. *Let $h = G\mu$. For any $B \in \mathscr{B}$ and $x \in E_h$*

$$(2.2) \qquad {}_hP^x\{X_{\zeta-} \in B\} = h(x)^{-1} \int_B g_y(x)\mu(dy).$$

PROOF. By Proposition 1.3 and Theorem 1.7 in turn, the left-hand side of (2.2) equals

$$\frac{1}{h(x)} \int_E g_y(x)\,{}_yP^x\{X_{\zeta-} \in B\}\mu(dy) = \frac{1}{h(x)} \int_B g_y(x)\mu(dy). \quad \text{Q.E.D.}$$

If $A \subset E$, define the *first entrance* and *last exit* times of the left-continuous process X_{t-} by

$$(2.3) \qquad \begin{cases} S_A = \inf\{t > 0: X_{t-} \in A\}, \\ L_A = \sup\{t > 0: X_{t-} \in A\}, \end{cases}$$

with the convention that the empty set has supremum zero and infimum plus infinity. If A is Borel, S_A is a stopping time and L_A is co-optional. A familiar capacity argument shows that for a given initial distribution μ, there exists a sequence K_n of subcompacts of A such that with P^μ-probability one, $S_{K_n} \downarrow S_A$ and $L_{K_n} \uparrow L_A$. An equally familiar argument using hypothesis (L) shows that K_n can be chosen independent of μ.

Notice that if K is compact, L_K is finite (transiency) and $X_{L_K-} \in K$ on $\{L_K > 0\}$, since $t \to X_{t-}$ is left continuous.

For a Borel set A, define

$$p_A(x) = P^x\{S_A < \infty\} = P^x\{L_A > 0\}. \tag{2.4}$$

Then p_A is excessive and, by our above remarks,

$$p_A(x) = \sup\{p_K(x): K \subset A \text{ is compact}\}. \tag{2.5}$$

PROPOSITION 2.2. *If K is compact, p_K is a Green's potential supported by K.*

PROOF. By (2.4) and Proposition 1.8, if Y is the process killed at L_K, Y is a p_K-transform of X. Now if $L_K > 0$, $Y_{\zeta-} = X_{L_K-} \in K$, so (2.1) holds with $h = p_K$; hence by Proposition 2.0, $p_K = G\mu$ and $\mu(K^c) = 0$.

Note. A similar argument shows that the function $f_K(x) = P^x\{T_K < \infty\}$ is also a Green's potential, but without further hypotheses it need not be supported by K.

A Borel set $A \subset E$ is *polar* if $P^x\{T_A < \infty\} \equiv 0$ and is *left polar* if $P^x\{S_A < \infty\} \equiv 0$. Define N_∞ by

$$N_\infty = \{y \in E - N: {}_yP^x\{\zeta = \infty\} = 1, x \in E_{g_y}\}. \tag{2.6}$$

The set N_1 of y which are the poles of more than one potential is left polar by hypothesis. What about the set N of y which are poles of none? The following proposition tells us that it is small, so there are many poles.

PROPOSITION 2.3. *N and N_∞ are left polar.*

PROOF. It is enough to show that any subcompact K of N or N_∞ is left polar. By (2.2) and Proposition 2.2

$$p_K(x) = \int_{K-N} g_y(x)\mu(dy).$$

This vanishes if $K \subset N$. If $K \subset N_\infty$, $L_K < \infty$ by transience, so that the process killed at L_K has a finite lifetime and, by Proposition 1.8, is a p_K-transform of X. By Proposition 1.3, if $x \in E_{p_K}$, then

$$1 = {}_{p_K}P^x\{\zeta < \infty\} = \frac{1}{p_K(x)} \int_K g_y(x)\, {}_yP^x\{\zeta < \infty\}\mu(dy).$$

But the integrand vanishes if $y \in K$, which is a contradiction unless $E_{p_K} = \emptyset$, i.e., $p_K \equiv 0$. Q.E.D.

We need two technical results on first hitting times. Let $A \in \mathscr{B}$ and define

$$f(x) = E^x\{e^{-S_A}\}, \qquad S_n = \inf\{t: f(X_t) > 1 - 1/n\}. \tag{2.7}$$

LEMMA 2.4. *$S_n \uparrow S_A$, and $S_n < S_A$ P^x-a.s. on $\{X_{S_A-} \in A\}$, $\forall x$.*

PROOF. Let $S = \lim S_n$. Certainly $S_A \leqq S$ if $S = \infty$, and if $S < \infty$, $f(X_{S_n}) \geqq 1-1/n$ by right continuity. But $f(X_{S_n}) = E\{\exp(-S_A \circ \theta_{S_n}) | \mathfrak{F}_{S_n}\}$, so $S_A \circ \theta_{S_n} \to 0$ in probability, and we conclude that $S_A \leqq S$.

Conversely, if $S_A < \infty$ and $X_{S_A-} \notin A$, S_A is a limit from the right of $\{t: X_{t-} \in A\}$.

By the strong Markov property, $f(X_{S_A}) = 1$ a.s.; hence $S_n \leqq S_A \ \forall n$. On the other hand, consider

$$T = \begin{cases} S_A & \text{if } X_{S_A-} \in A, \\ \infty & \text{otherwise.} \end{cases}$$

Note that $\{T < \infty\} \in \mathfrak{F}_{T-}$; hence by [**3**, Théorème 49, p. 61] T is predictable. Let (T_n) announce T. Then on $\{T < \infty\}$, $S_A \circ \theta_{T_n} \leqq T - T_n \to 0$. Thus

$$f(X_{T_n}) \geqq E\{e^{-(T-T_n)} \mid \mathfrak{F}_{T_n}\} \to 1$$

a.s. on $\{T < \infty\}$. But $T_n < T$ for all n, so that for each m, $S_m < T$. Thus we have $S \leqq T$, hence $S = T$ and we are done.

LEMMA 2.5. *Let $A \in \mathscr{B}$. There exist $\mathfrak{F}^0$-measurable stopping times S and T and co-optional L and M such that for all x*

$$P^x\{S = S_A,\ T = T_A,\ L = L_A \text{ and } M = M_A\} = 1.$$

PROOF. Let $\hat{S}_n = \inf\{t\colon t \in \mathbf{Q},\ f(X_t) > 1 - 1/n\}$. By right continuity of $f(X_t)$, $\hat{S}_n = S_n$ P^μ-a.s. for any μ, where S_n and f are defined in (2.7). f, being 1-excessive, is Borel measurable, so $\hat{S}_n$ is $\mathfrak{F}^0$-measurable. If $S \overset{\text{def}}{=} \sup S_n$, $P^\mu\{S = S_A\} = 1$ by Lemma 2.4. Letting $L = \sup\{t \in \mathbf{Q}\colon S \circ \theta_t < \infty\}$, it follows that $P^\mu\{L = L_A\} = 1$. The proof for T_A and M_A is similar. Q.E.D.

DEFINITION. A Borel set A is *cothin* at $y \in E - N$ if A is g_y-thin.

PROPOSITION 2.6. *Let A be Borel. Then so is the set of points at which A is cothin. If*

$$\Gamma_n(A) = \{x\colon E^x\{e^{-S_A}\} > 1 - 1/n\}, \tag{2.8}$$

the set of points at which $\Gamma_n(A)$ is cothin is left polar.

PROOF. Let μ_0 be a reference measure, and let M be the $\mathfrak{F}^0$-measurable co-optional time which equals M_A. Note that ${}_yP^{\mu_0}\{M = M_A\} = 1$ for all y. Since ${}_yP^{\mu_0}\{M < \zeta\}$ is Borel in y by Proposition A6, we have $\{y\colon {}_yP^{\mu_0}\{M < \zeta\} > 0\} \in \mathscr{B}$. This is exactly the set at which A is cothin.

Now let $R = \{y \in A\colon \Gamma_n(A) \text{ is cothin at } y\}$. Then $\Gamma_n(A)$, and hence R, is in $\mathscr{B}$. Let $K \subset R$ be compact. By Lemma 2.4, $T_{\Gamma_n(K)} \leqq S_K$ P^x-a.s., which follows since $L_K = \inf\{t \in \mathbf{Q}\colon S_K \circ \theta_t = \infty\}$ and similarly for $M_{\Gamma_n(K)}$. The process killed at L_K is the p_K-transform of X, so that for $x \in E_{p_K}$

$$0 = {}_{p_K}P^x\{L_K < \zeta\} \geqq {}_{p_K}P^x\{M_{\Gamma_n(K)} < \zeta\} = p_K(x)^{-1} \int_K g_y(x)\, {}_yP^x\{M_{\Gamma_n(K)} < \zeta\}\, \mu(dy).$$

But $\Gamma_n(K) \subset \Gamma_n(A)$ and $\Gamma_n(A)$ is cothin at each $y \in K$, so the probability on the right-hand side is one, a contradiction unless $E_{p_K} = \varnothing$. Thus K is left polar. K was arbitrary, so R must be left polar, too. Q.E.D.

A compact K is said to have *capacity zero* if there exists no nontrivial bounded Green's potential with support in K. A Borel set A has capacity zero if all subcompacts have capacity zero.

THEOREM 2.7. *A Borel set A is left polar iff $A - N_\infty$ is of capacity zero.*

REMARK. The reason for excluding N_∞ can be seen by considering Brownian motion on $\boldsymbol{R}^3$, where the point at infinity is left polar, but not of capacity zero.

PROOF. If A is not left polar, it is not of capacity zero, for $\exists$ a nonleft polar compact $K \subset A$, and p_K is a nontrivial Green's potential with support in K by Proposition 2.2.

Conversely, if $K \subset A - N_\infty$ is compact and $h = G\nu$ is a bounded Green's potential with support in K, then for $x \in E_h$, ${}_hP^x\{X_{\zeta-} \in K, \zeta < \infty\} = 1$, so ${}_hP^x\{S_K \leqq \zeta < \infty\} = 1$. If ${}_hP^x\{S_K < \zeta\} > 0$, then

$$0 < {}_hP^x\{S_K < \zeta\} = h(x)^{-1} E^x\{h(X_{S_K})\};$$

in particular, $P^x\{S_K < \infty\} > 0$, so K is not left polar. However, we may have ${}_hP^x\{S_K = \zeta\} = 1$, all $x \in E_h$. Let (G_n) be a decreasing family of open neighborhoods of K such that $\bar{G}_n \downarrow K$. Then ${}_hP^x\{T_{G_n} < S_K, T_{G_n} \uparrow S_K\} = 1$.

Choose t such that ${}_hP^x\{T_{G_n} \leqq t, \forall n\} = p > 0$ and notice that $\{T_{G_n} \leqq t\} \in \mathfrak{F}_{T_{G_n}}$, so that by Theorem 1.2,

$$\begin{aligned} 0 < p \leqq {}_hP^x\{T_{G_n} < \zeta, T_{G_n} \leqq t\} &= h(x)^{-1} E^x\{T_{G_n} \leqq t; h(X_{T_{G_n}})\} \\ &\leqq \|h\| h(x)^{-1} P^x\{T_{G_n} \leqq t\}, \end{aligned} \tag{2.9}$$

where $\|h\| = \sup h$. Let $n \to \infty$; $\lim X_{T_{G_n}} \in K$ so $S_K \leqq \lim T_{G_n}$ and we can go the limit in (2.9):

$$P^x\{S_K \leqq t\} \geqq \|h\|^{-1} h(x)p > 0,$$

so K is not left polar. We already know N_∞ is left polar (Proposition 2.3), so we are done.

LEMMA 2.8. *Let $G\mu$ be a bounded Green's potential of compact support. Then* $\lim_{t\to\infty} G\mu(X_t) = 0$ P^x-a.s., $\forall x$.

PROOF. Let V be a relatively compact neighborhood of supp(μ). If $y \in V$, $g_y(x) = R_V g_y(x) = E^x\{g_y(X_{T_V})\}$. Thus

$$G\mu(x) = \int_V E^x\{g_y(X_{T_V})\}\mu(dy) = E^x\{G\mu(X_{T_V})\} \leqq \|G\mu\| P^x\{T_V < \infty\}.$$

Then

$$G\mu(X_t) \leqq \|G\mu\| P^x\{T_V \circ \theta_t < \infty\},$$

which goes to zero by transience. Q.E.D.

This brings us to the two results we have been leading up to, which will give a purely analytic characterization of the hitting probabilities p_A.

THEOREM 2.9. *Let $A \subset E$ be Borel. Then*

$$p_A(x) = \sup\{G\mu(x) : G\mu \leqq 1, \text{supp}(\mu) \subset A - N_\infty \textit{ is compact}\}. \tag{2.10}$$

PROOF. By Proposition 2.2, the left-hand side of (2.7) is dominated by the right. We must show "$\geqq$". It is enough to do this when A is compact in $E - N_\infty$. Let Γ_n be defined by (2.8). If $G\mu \leqq 1$, μ cannot charge a set of capacity zero (Proposition 2.8); hence it does not charge the set $B = \{y : \Gamma_n \text{ is cothin at } y\}$. Suppose that $y \in A - B$ and $x \in E_{g_y}$. By Propositions 1.4 and 1.6:

$$g_y(x) = R_{\Gamma_n} g_y(x) = E^x\{g_y(X_{T_n})\},$$

where T_n is the first hit of Γ_n. Thus

$$(2.11)\quad G\mu(x) = \int_{A-B} g_y(x)\mu\,(dy) = \int_{A-B} E^x\{g_y(X_{T_n})\}\mu\,(dy) = E^x\{G\mu(X_{T_n})\}.$$

Let $n \to \infty$; $T_n \uparrow S_A$ by Lemma 2.4 and, since $\mu(N_\infty) = 0$, $G\mu(X_t) \to 0$ as $t \to \infty$ (Lemma 2.8). Thus $G\mu(X_{T_n}) \to 0$ on $\{S_A = \infty\}$; as $G\mu \leqq 1$ we conclude

$$G\mu(x) \leqq P^x\{S_A < \infty\} = p_A(x). \quad \text{Q.E.D.}$$

Let us notice that if $y \in (N \cup N_\infty)^c$, the g_y-transform of X is transient (for its lifetime is finite), has a left limit at ζ (for $X_{\zeta-} = y$), satisfies hypothesis (L), and, by Theorem B3 of Appendix B, it satisfies hypothesis (G). In short, everything above is valid for the g_y-transform as well as for the original process. Moreover, the Green's function ${}_yg$ has the form

$${}_yg_z(x) = \begin{cases} g_y(x)^{-1}\, g_z(x) & \text{if } x \in E_{g_y}, \\ 0 & \text{otherwise.} \end{cases}$$

Let ${}_yp_A(x) = {}_yP^x\{S_A < \infty\}$. Applying Theorem 2.9, we get

COROLLARY 2.10. *Let $A \subset E$ be Borel and $y \in E - N$. Then*

$$(2.12)\qquad {}_yp_A(x) = \frac{1}{g_y(x)} \sup\{G\mu(x) : G\mu \leqq g_y \text{ and } \operatorname{supp}(\mu) \text{ is compact in } A - N_\infty\}.$$

3. The cofine topology. We are now able to define the cofine topology in our setting, and to show that it satisfies the conditions set forth in **[20]**. Actually, it is misleading to speak of "the" cofine topology, for according to **[18]**, one can modify such a topology on a left polar set and still have a cofine topology. The definition given here would appear to be canonical, although there is a source of ambiguity hidden in the definition of the Green's function in Appendix A, which depended on the choice of a particular cap of the cone of excessive functions.

We combine the exceptional sets N, N_1, and N_∞ into one set, which we call N.

DEFINITION. A Borel set A is *elusive* at a point $y \in E - N$ if ${}_yp_A(x) < 1$ for some $x \in E_{g_y}$. We say A is elusive at a point $y \in N$ if y is not in the ordinary closure of A.

Our definition of elusiveness at points of the exceptional set N may seem arbitrary, as indeed it is, but N is left polar (Proposition 2.3) and can safely be ignored in what follows. Notice that if $y \in A - N$ then A is not elusive at y (for ${}_yP^x\{\zeta < \infty, X_\zeta = y \in A\} = 1$ for all $x \in E_{g_y}$), so that the idea of elusiveness is of interest only when $y \notin A$.

Corollary 2.10 gives a potential-theoretic interpretation of elusive sets; here is a probabilistic explanation.

PROPOSITION 3.1. *Let $y \in E - N$ and let $A \in \mathscr{B}$ such that $y \notin A$. Then A is elusive at y iff*

$$(3.1)\qquad {}_yP^x\{L_A < \zeta\} = 1, \quad \textit{all } x \in E_{g_y}.$$

PROOF. We remark that $S_A = \infty$ iff $L_A = 0$, so for $x \in E_{g_y}$

$$(3.2)\qquad 1 - {}_yp_A(x) = {}_yP^x\{S_A = \infty\} = {}_yP^x\{L_A = 0\} \leqq {}_yP^x\{L_A < \zeta\}.$$

Now $\{L_A < \zeta\}$ is invariant, so the right-hand side of (3.2) is either zero or one.

If A is elusive, the left-hand side is strictly positive for some x, which implies that (3.1) holds. Conversely, if (3.1) holds, there exists t such that

$$0 < {}_yP^x\{L_A \leqq t < \zeta\} = {}_yP^x\{S_A \circ \theta_t = \infty, \zeta > 0\} = {}_yE^x\{1 - {}_yp_A(X_t)\}.$$

Thus ${}_yp_A \not\equiv 1$, and A is elusive at y. Q.E.D.

To rephrase (3.1), if A is elusive at y, then X_{t-} is in $E - A$ for all t sufficiently close to ζ. This suggests the following definition.

DEFINITION. A Borel set $V \subset E$ is a *cofine neighborhood* of $y \in E$ if $y \in V$ and if $E - V$ is elusive at y.

It is clear from (3.1) that the intersection of two cofine neighborhoods is a cofine neighborhood. Thus the cofine neighborhoods form a neighborhood base for a topology on E, called the *cofine topology*. We extend this to $E \cup \delta$ as usual, by making δ an isolated point. It is clear from Theorem 1.7 that any ordinary neighborhood of y is a cofine neighborhood. Thus the cofine topology is finer than the ordinary topology.

PROPOSITION 3.2. *Let A be Borel. Then the cofine closure of A is also Borel.*

PROOF. Let μ_0 be a reference measure. Let L be an $\mathfrak{F}^0$-measurable time which is ${}_yP^{\mu_0}$-a.s. equal to L_A, for each y (see Lemma 2.5). In view of Proposition 3.1, the cofine closure of A is

$$A \cup (N \cap \bar{A}) \cup \{y \in E - N: {}_yP^{\mu_0}\{L = \zeta\} > 0\}.$$

This is Borel by Proposition A6. Q.E.D.

We can now prove that the name "cofine topology" is appropriate. More specifically, we have:

THEOREM 3.3. *The cofine topology defined above is a cofine topology in the sense of* [**20**], *i.e., it satisfies hypothesis* (CF).

PROOF. First, we verify (CF1) (see the introduction). Let f be a bounded Borel function which is cofine continuous, except possibly at a left polar set. To show $t \to f(X_{t-})$ is left continuous, it is enough to show it is left continuous on $(0, L_K]$, where K is any compact set. Let $\tilde{X}$ be the right continuous process $\{\tilde{X}_t = X_{L_K-t-}, 0 \leqq t \leqq L_K\}$. It is then sufficient to show that $f(\tilde{X}_t)$ is *right* continuous, and even enough to show $f(\tilde{X}_t)$ is right continuous at a given stopping time T [**3**, Théorème 28, p. 82]. But $(L_K - T) \vee 0$ is co-optional for the original process [**11**, p. 37], so, finally, it is enough to show that for any co-optional time L,

$$\lim_{s \uparrow L} f(X_{s-}) = f(X_{L-}) \quad P^x\text{-a.s., all } x.$$

Let $y \in E - N$ be such that f is cofine continuous at y. The set $J_n = \{x: |f(x) - f(y)| > 1/n\}$ is then elusive at y; hence ${}_yP^x\{L_{J_n} < \zeta\} = 1$, $x \in E_{g_y}$, by Proposition 3.1. It follows that $|f(X_{s-}) - f(y)| < 1/n$ for s sufficiently close to ζ. As n is arbitrary,

$$(3.3) \qquad {}_yP^x\left\{\lim_{s \uparrow \zeta} f(X_{s-}) = f(y)\right\} = 1.$$

If L is a finite co-optional time, the process killed at L is an h-transform, with

$h(x) = P^x\{L > 0\}$ (Theorem 1.8). Since X_{L-} exists, Proposition A8 tells us that h is a Green's potential, say $h = G\mu$. Noting that ${}_yP^x\{X_{\zeta-} = y\} = 1$, by Proposition 1.3, we have, if $\Lambda = \{\lim_{s\uparrow L} f(X_{s-}) = f(X_{L-})\}$

$$(3.4)\qquad \begin{aligned} P^x\{\Lambda \mid L > 0\} &= \frac{1}{h(x)}\int g_y(x)\,{}_yP^x\Big\{\lim_{s\uparrow\zeta} f(X_{s-}) = f(y)\Big\}\mu(dy) \\ &= \frac{1}{h(x)}\int g_y(x)\mu(dy) = 1, \end{aligned}$$

where the penultimate equality follows from (3.3).

Passing to (CF2), suppose f is a bounded Borel function for which $s \to f(X_{s-})$ has left limits P^x-a.s. for all x. If y is not isolated in the cofine topology, define

$$\bar f(x) = \operatorname*{cofine\ lim\ sup}_{x\to y,\, x\neq y} f(x) \quad\text{and}\quad \underline{f}(x) = \operatorname*{cofine\ lim\ inf}_{x\to y,\, x\neq y} f(y),$$

and, if y is isolated, put $\bar f(y) = \underline{f}(y) = f(y)$. Both $\bar f$ and $\underline{f}$ are Borel by Proposition 3.2. The set on which f is not cofine continuous is contained in $A = \{y : \bar f(y) > \underline{f}(y)\}$. If K is a compact subset of A, p_K is a Green's potential, say $p_K = G\mu$, and $\operatorname{supp}(\mu) \subset K$. The process killed at L_K being a p_K-transform, if $x \in E_{p_K}$

$$(3.5)\qquad \begin{aligned} P^x\Big\{\overline{\lim_{s\uparrow L_K}} f(X_{s-}) > \underline{\lim_{s\uparrow L_K}} f(X_{s-}) \,\Big|\, L_K > 0\Big\} &= {}_{p_K}P^x\Big\{\overline{\lim_{s\uparrow\zeta}} f(X_{s-}) > \underline{\lim_{s\uparrow\zeta}} f(X_{s-})\Big\} \\ &= \frac{1}{p_K(x)}\int_K g_y(x)\,{}_yP^x\Big\{\overline{\lim_{s\uparrow\zeta}} f(X_{s-}) > \underline{\lim_{s\uparrow\zeta}} f(X_{s-})\Big\}\mu(dy). \end{aligned}$$

But for $y \in K \cap E - N$ and $\varepsilon > 0$ both $\{x : f(x) > \bar f(y) - \varepsilon\}$ and $\{x : f(x) < \underline{f}(y) + \varepsilon\}$ are not elusive at y; by Proposition 3.1 the g_y-transform will enter both arbitrarily near to ζ. In other words,

$$(3.6)\qquad {}_yP^x\Big\{\overline{\lim_{s\uparrow\zeta}} f(X_{s-}) > \bar f(y) - \varepsilon,\ \underline{\lim_{s\uparrow\zeta}} f(X_{s-}) < \underline{f}(y) + \varepsilon\Big\} = 1.$$

Apply this to (3.5). The right-hand side of (3.5) is then

$$= \frac{1}{p_K(x)}\int_K g_y(x)\mu(dy) = 1.$$

But now $s \to f(X_{s-})$ has left limits P^x-a.s., so $P^x\{0 < L_K < \infty : \overline{\lim}_{s\uparrow\zeta} f(X_{s-}) > \underline{\lim}_{s\uparrow\zeta} f(X_{s-})\} = 0$. This contradicts (3.5) unless $E_{p_K} = \varnothing$, i.e., unless K is left polar. K is arbitrary, so it follows that A is left polar. Thus the set at which f fails to have a cofine limit is left polar.

Now the set $\{x : \bar f(x) < f(x) - \varepsilon\}$ can have no cofine accumulation points, from which it follows that $\bar f$ is cofine continuous except at A. By (CF1), which we have just verified, both $s \to \bar f(X_{s-})$ and $s \to \underline{f}(X_{s-})$ are left continuous.

Finally, consider the two left continuous processes $Y_t = \lim_{s\uparrow t} f(X_{s-})$ and $Z_t = \bar f(X_{t-})$. We claim they are indistinguishable. By the argument beginning this proof, it is enough to show equality at an arbitrary finite co-optional time L. Let $h = G\mu$ be the Green's potential $P^x\{L > 0\}$. If $\Lambda = \{\lim_{s\uparrow\zeta} f(X_{s-}) = \bar f(X_{\zeta-})\}$, then, because the process killed at L is an h-transform,

$$(3.7)\qquad P\{Y_L = Z_L \mid L > 0\} = {}_hP^x\{\Lambda\} = \frac{1}{h(x)}\int g_y(x)\,{}_yP^x\{\Lambda\}\mu(dy).$$

Since the set $A = \{x : \bar f(x) > \underline{f}(x)\}$ is left polar, we may assume that $\bar f(y) =$

cofine $\lim_{x \to y,\, x \neq y} f(x)$, which means that the set $\{x \neq y : |f(x) - \bar{f}(y)| > \varepsilon\}$ is elusive at y, for any $\varepsilon > 0$. It follows that

$$_yP^x\Big\{ \lim_{s \uparrow \zeta,\, X_{s-} \neq y} f(X_{s-}) = \bar{f}(y) \Big\} = 1;$$

but we know that $\lim_{s \uparrow \zeta} f(X_{s-})$ exists, so if y is not in a left polar set, $_yP^x\{\Lambda\} = 1$, $x \in E_{g_y}$; hence the last term in (3.7) is one, and we are done.

4. Two examples. We can get some insight into the meaning of hypothesis (G) by considering some cases where it is violated, and to see why, in these cases, there is no cofine topology. The classical processes such as Brownian motion and the stable processes have a cofine topology and also satisfy hypothesis (G).

Both examples are derived from uniform motion, so the reader may want to check the following fact for himself. Consider uniform motion to the right on $(-\infty, \infty)$. Then the function $f(x) = I_{(-\infty, y)}(x)$ is the unique (up to proportionality) minimal excessive function with pole y.

EXAMPLE 1. Uniform motion to the right on the lazy crotch (see the design).

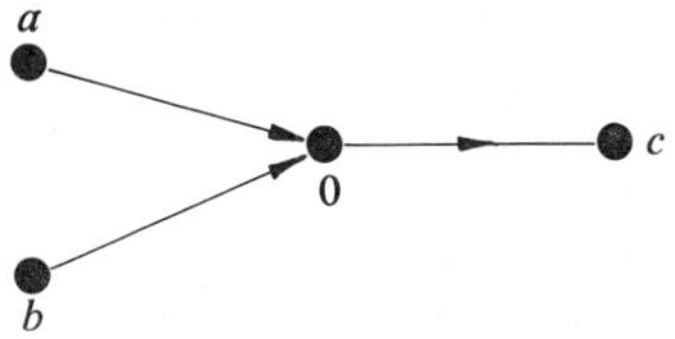

This is the standard example in which no cofine topology exists. Hypothesis (G) does not hold: indeed, the functions

$$f_a(x) = \begin{cases} 1 & \text{if } x \in [a, 0), \\ 0 & \text{otherwise,} \end{cases} \quad \text{and} \quad f_b(x) = \begin{cases} 1 & \text{if } x \in [b, 0), \\ 0 & \text{otherwise,} \end{cases}$$

are both minimal excessive functions with pole 0. In this case, the problem comes from the fact that there are two distinct ways to approach the origin, and there is one excessive function corresponding to each way.

EXAMPLE 2. Let $E = (-\infty, 0]$, let X be uniform motion on $(-\infty, 0]$, and make 0 an exponential holding point. The process is killed upon leaving zero. Once again there are two minimal excessive functions with pole 0:

$$f(x) \equiv I_{(-\infty, 0)}(x) \quad \text{and} \quad g(x) \equiv 1.$$

The reason for the lack of a cofine topology is less evident than before: it is (CF2) that breaks down. Indeed, if we start with f above, then if there were a cofine topology, the only possibilities for $\bar{f}$ would be $\bar{f} = f$ or $\bar{f} = g$. If $\bar{f} = f$, then $t \to \bar{f}(X_{t-})$ would fail to be left continuous at the first hit T_0 of zero; if $\bar{f} = g$, $\bar{f}(X_{t-})$ would be continuous, but we would not have $\bar{f}(X_{t-}) = \lim_{s \uparrow t} f(X_{s-})$ if $T_0 < t < \zeta$.

A more intuitive way of seeing that no cofine topology exists is to notice that if it did, the reversed process would be a right continuous strong Markov process, and its fine topology would be the cofine topology we are looking for. But the reversed process is a continuous process which has an exponential holding point 0, and cannot be strong Markov (consider the first time the process leaves zero).

5. Two related topologies. There are several topologies which are related to the topology defined in §3. We will consider two here, the topology defined by cothin sets and Getoor's d-topology. We keep the notation and hypotheses of §2.

Malte Sieveking suggested [personal correspondence] that under hypothesis (G), one could define a topology as follows: say a Borel set V is a neighborhood of $y \in E - N$ in the $\mathcal{S}$-topology if $y \in V$ and if V^c is cothin at y (see §2 and Proposition 1.6), and V is a neighborhood of $y \in N$ if it is an ordinary neighborhood of y. These neighborhoods form a base for a topology, which we will call the $\mathcal{S}$-topology.

To compare the $\mathcal{S}$-topology with the cofine topology, remark that if M_A and L_A are respectively the last exits of X_t and X_{t-} from A, then by Proposition 1.6, V is an $\mathcal{S}$-neighborhood of y iff $y \in V$ and

$$(5.1) \qquad {}_yP^x\{M_{V^c} < \zeta, \zeta > 0\} = 0,$$

while V is a cofine neighborhood iff $y \in V$ and

$$(5.2) \qquad {}_yP^x\{L_{V^c} < \zeta, \zeta > 0\} = 0.$$

The difference between the two only involves interchanging X_t and X_{t-}, but it is not hard to find examples in which the $\mathcal{S}$-topology is not a cofine topology. However, the two topologies coincide in case X is continuous, and in fact they coincide in the case of standard processes in duality, as we will see.

Recall that X is said to satsify Hunt's hypothesis (B) if, for each semipolar B, $P^x\{\exists\, t: X_t \in B \text{ and } X_{t-} \neq \hat{X}_t\} = 0$, all $x \in E$.

THEOREM 5.1. *Suppose that X is quasi-left continuous, and satisfies Hunt's hypothesis* (B). *Then the $\mathcal{S}$-topology and the cofine topology coincide.*

REMARK. If X and $\hat{X}$ are Hunt processes in duality, both satisfy hypothesis (B) **[18]**.

LEMMA 5.2. *Let $A \in \mathcal{B}$. If X is quasi-left continuous, $P^x\{T_A \leqq S_A\} = 1$. If X satisfies hypothesis* (B), $P^x\{S_A \leqq T_A\} = 1$.

PROOF. By the usual approximation argument, it is enough to consider the case where A is compact. Suppose X is quasi-left continuous and let $f(x) = E^x\{e^{-S_A}\}$. S_A is predictable on $\{X_{S_A-} \in A\}$ (Lemma 2.4); hence $X_{S_A-} = X_{S_A} \in A$, while on the set $\{X_{S_A-} \notin A\}$, we must have $f(X_{S_A}) = 1$ by the strong Markov property. But $\{x: f(x) = 1\} \subset A$. Thus in either case $T_A \leqq S_A$.

The second conclusion seems curiously more difficult. Suppose X satisfies hypothesis (B) and let U_A be the terminal time

$$U_A = \inf\{t: X_{t-} \in A \quad\text{and}\quad X_{t-} = X_t\},$$

and let $h(x) = E^x\{e^{-U_A}\}$, and let $A' = \{x: h(x) = 1\}$. Suppose we can show that $A - A'$ is semipolar. Since A is compact, $X_{T_A} \in A$. If $X_{T_A} \in A'$, then, since $S_A \leqq U_A$, $S_A \circ \theta_{T_A} = 0$ a.s., so $S_A \leqq T_A$. On the other hand, by hypothesis (B), $P^x\{X_{T_A} \in A - A', X_{T_A-} \neq X_{T_A}\} = 0$, and again $S_A \leqq T_A$.

It remains to show that $A - A'$ is semipolar. Consider the set $B = A \cap \{x: f(x) \leqq 1 - \varepsilon\}$, and let $U^1 = U_B$, $U^{n+1} = U^n + U_B \circ \theta_{U^n}$. A is compact and B is fine closed, hence $X_{U^n} \in B$ if $U^n < \infty$. Since $B \subset A$, $U_B \geqq U_A$, so on $\{U^n < \infty\}$,

$$E\{e^{-(U^{n+1}-U^n)} \mid \mathfrak{F}_{U^n}\} \leqq 1 - \varepsilon.$$

It follows by a standard argument that $\lim U^n = \infty$. But consider the interval (U^n, U^{n+1}). There cannot be a t in this interval for which $X_t \in B$ and $X_{t-} = X_t$. Since X_t is continuous except on a countable set, X_t can be in B for at most countably many $t \in (U^n, U^{n+1})$, and hence for at most countably many $t \geqq 0$. By a theorem of Dellacherie [4], B is semipolar. Then so is $A - A'$, which is a countable union of such sets. Q.E.D.

PROOF OF THEOREM 5.1. Let $V \in \mathscr{B}$ contain y. Since $y \notin V^c$, ${}_yP^x\{S_{V^c} = \zeta \text{ or } T_{V^c} = \zeta\} = 0$. It follows from this and the lemma that ${}_yP^x\{L_{V^c} = M_{V^c}\} = 1$ for $x \in E_{g_y}$, since

$$L_A = \sup\{t \in Q\colon S_{V^c} \circ \theta_t < \infty\} \quad \text{and} \quad M_A = \sup\{t \in Q, T_{V^c} \circ \theta_t < \infty\}.$$

Thus (5.1) and (5.2) reduce to the same condition. Q.E.D.

A second related topology, called the d-topology, was introduced by Getoor in [8] in order to study regular excessive functions. This can be defined without duality or hypothesis (G), so there is no reason to suppose that it should coincide with the cofine topology, but it is in fact related in an interesting way: the neighborhood bases of the two topologies may differ at each point; nevertheless, a d-open set is a cofine neighborhood of quasi every one of its points.

We say a Borel set D is a *d-set* if, for each initial measure μ and each increasing sequence $\{T_n\}$ of stopping times with the property that $X_{T_n} \in D$ P^μ-a.s. on $\{T_n < \infty\}$, $X_T \in D$, where $T = \lim T_n$. The complements of d-sets form a base for a topology called the *d-topology*.

PROPOSITION 5.3. *Suppose X is a Hunt process satisfying hypothesis* (B). *Then the d-topology is finer than the cofine topology. Conversely, if A is Borel measurable and open in the d-topology, it is a cofine neighborhood of all but a left polar set of $y \in A$.*

PROOF. Let $D \in \mathscr{B}$ and let K be compact in $E - D$. Let $M_D(t) = \sup\{s < t\colon X_s \in D\}$ be the last exit before t, and define a predictable set $\Lambda_{D,K}$ by

$$\Lambda_{D,K} = \{(t, \omega)\colon t > 0, X_{t-} \in K \text{ and } M_D(t) = t\}.$$

We claim D is a d-set iff $\Lambda_{D,K}$ is evanescent for all compacts $K \subset D^c$, and all initial distributions. For if $\Lambda_{D,K}$ is not P^x-evanescent, by the section theorem [3, p. 71] there exists a predictable time T with $[T] \subset \Lambda_{D,K}$, and $P^x\{T < \infty\} > 0$. Let (T_n) announce T. Now $M_D(T) = T$ on $\{T < \infty\}$, so the process hits D infinitely often on the interval (T_n, T). Using the section theorem again, we can find a stopping time T_n' such that $X_{T_n'} \in D$ a.s. on $\{T < \infty\}$ and $T_n \leqq T_n' < T$. Since T is predictable, $X_T = X_{T-} \in K$. But $X_{T_n'} \in D$ and $X_T \in D^c$ means that D is not a d-set. Conversely, if D is not a d-set, there exist $x \in E$, stopping times $T_n \uparrow T$, and a compact $K \subset E - D$ such that $P^x\{X_{T_n} \in D, \forall\, n \text{ and } X_T \in K\} > 0$. By quasi-left continuity, $\lim X_{T_n} = X_T \in K$, so

$$P^x\{X_{T-} \in K, M_D(T) = T\} > 0,$$

and $\Lambda_{D,K}$ is not evanescent.

Now suppose A is cofine open and let $K \subset A$ be compact. Let $L(\omega) = \sup\{t\colon (t, \omega) \in \Lambda_{A^c,K}\}$ be the end of $\Lambda_{A^c,K}$. Notice that L is co-optional. By an argument often used above, the process killed at L is an h-transform, with $h(x) = P^x\{L > 0\} = G\mu$, where $\text{supp}(\mu) \subset K$. Thus if $x \in E_h$,

$$P^x\{M_{A^c} = L > 0\} = \int_K g_y(x)\,{}_yP^x\{M_{A^c} = \zeta\}\mu(dy) = 0$$

since A^c is elusive, hence cothin (Theorem 5.1), at all $y \in K$.

Finally, suppose A^c is a d-set and let K be a compact subset of those $y \in A$ for which A^c is not elusive (and hence not cothin). The process killed at L_K is a p_K-transform, and $p_K = G\mu$ with $\operatorname{supp}(\mu) \subset K$, so

$$\begin{aligned} P^x\{M_{A^c}(L_K) = L_K > 0\} &= \int_K g_y(x)\,{}_yP^x\{M_{A^c} = \zeta\}\mu(dy) \\ &= \int_K g_y(x)\mu(dy) = p_K(x), \end{aligned}$$

since A is not a cofine neighborhood of any $y \in K$. Thus if $L_K > 0$, then $X_{L_K-} \in K$ and $M_{A^c}(L_K) = L_K$, i.e., $[L] \subset \Lambda_{A^c,K}$. Therefore $\Lambda_{A^c,K}$ is not evanescent unless K is left polar. Q.E.D.

Interestingly enough, it may happen that the d-topology is strictly finer than the cofine topology at every point, even in the best of cases. For instance, consider Brownian motion in the plane. If x is any point and D a line passing through x, D, being closed, is a d-set. Since points are polar, $D' = D - \{x\}$ is still a d-set. Thus $\boldsymbol{R}^2 - D'$ is a neighborhood of x in the d-topology, but it is not a cofine neighborhood. Indeed, the fine and cofine topology coincide in this case, and it is well known that Brownian motion from x will encounter D' immediately.

APPENDIX A

The Green's function. We are going to construct a Green's function and derive some of its elementary properties, starting from the hypotheses of §2. We note that this has been done in the case of axiomatic potential theory by R. M. Hervé **[9]**, and recently extended by K. Janssen **[10]** to harmonic spaces; R. Duncan **[7]** has constructed a Green's function under more general hypotheses, which unfortunately do not quite cover our needs. Thus, as in §2, X is a right continuous strong Markov process having left limits which satisfies hypotheses (G) and (L). We do not assume X is transient. The Borel field of E will be denoted by $\mathscr{B}$, and μ_0 denotes a fixed regular reference probability measure.

Let $\mathscr{E}$ be the cone of excessive functions. By a theorem of Mokobodzki **[15]**, $\mathscr{E}$ is a union of caps C_f of the form

$$C_f = \{v \in \mathscr{E} : \langle v, f \rangle \leqq 1\}$$

where f is Borel and strictly positive on E and

$$\langle v, f \rangle = \int_E v(x)f(x)\mu_0(dx).$$

Furthermore, there exists a function g such that $g(x) = 0$ iff $f(x) = 0$, and such that if C_f is given the topology induced by $L^1(g \cdot \mu_0)$, it is compact and metrizable. Let $\mathscr{C}$ be the Borel field of C_f.

LEMMA A1. *If $K(v, x; dy)$ is a positive kernel from $C_f \times E$ to E, then $(v, x) \to Kv(x)$ is $\mathscr{C} \times \mathscr{B}$-measurable.*

PROOF. This is clear if K is of the form $K(v, x; dy) = p(v)q(x)r(y)\mu_0(dy)$, where p, q, and r are bounded and measurable, since then

$$Kv(x) = p(v)q(x)\langle v, r\rangle,$$

and $v \to \langle v, r\rangle$ is a continuous linear functional. This extends to K of the form $K(v, x; dy) = k(v, x, y)\mu_0(dy)$ by the monotone class theorem. In general, let

$$K_\lambda(v, x; \cdot) = \lambda \int K(v, x; dz)R_\lambda(z, \cdot),$$

where R_λ is the resolvent. Since $R_\lambda(z, \cdot) \ll \mu_0$ (μ_0 is a reference measure), K_λ is of the above form, so $(v, x) \to K_\lambda v(x)$ is $\mathscr{C} \times \mathscr{B}$-measurable. As v is excessive, $Kv(x) = \lim_{\lambda\to\infty} K_\lambda v(x)$. Q.E.D.

It follows that $(v, x) \to P_t v(x)$ and $(v, x) \to v(x)$ are $\mathscr{C} \times \mathscr{B}$-measurable (take $K(x, \cdot) = \delta_x(\cdot)$), as is $(v, x) \to {}_vP_t f(x)$ for a positive Borel f.

Let Ω be the space of functions from $[0, \infty)$ to $E \cup \partial$ which admit a lifetime ζ, are right continuous on $[0, \infty)$, and have left limits except possibly at ζ. Let X be defined canonically on Ω, and let $\mathfrak{F}^0$ and $\mathfrak{F}_t^0$ be the natural (i.e., uncompleted) fields.

PROPOSITION A2. *If* $\Lambda \in \mathfrak{F}^0$, *then* $(v, x) \to {}_vP^x\{\Lambda\}$ *is* $\mathscr{C} \times \mathscr{B}$*-measurable.*

PROOF. This is true by the above remark if $\Lambda = \{X_t \in A\}$ for some Borel set A, and for Λ of the form $\{X_{t_1} \in A_1, \cdots, X_{t_n} \in A_n\}$ by induction. It extends to all $\Lambda \in \mathfrak{F}^0$ by the monotone class theorem.

LEMMA A3. *For* $y \in E$, $(v, x, y) \to {}_vP^x\{X_{\zeta-} = y\}$ *is* $\mathscr{C} \times \mathscr{B} \times \mathscr{B}$*-measurable.*

PROOF. Let $\{A_{nj}\}_{j=1}^\infty$ be a partition of E into Borel sets of diameter bounded by $1/n$. For each j,

$$(v, x) \to {}_vP^x\{X_{\zeta-} \in A_{nj}\} \quad \text{is } \mathscr{C} \times \mathscr{B}\text{-measurable by A2.}$$

Let

$$u_n(v, x, y) = \sum_j I_{A_{nj}}(y)_vP^x\{X_{\zeta-} \in A_{nj}\}.$$

Then u_n is $\mathscr{C} \times \mathscr{B} \times \mathscr{B}$-measurable, and converges to ${}_vP^x\{X_{\zeta-} = y\}$ as $n \to \infty$. Q.E.D.

Consider the set $\mathscr{C}'_f = \{v \in C_f : \langle v, f\rangle = 1\}$. This is a compact, convex, metrizable set; hence the set ∂C_f of its extreme points is a G_δ **[16]**. These extreme points are minimal excessive functions.

Define a mapping ρ: $v \to$ pole of v, from those $v \in \partial C_f$ which have poles, to E. Recall that N_1 is the Borel set of hypothesis (G), so if $y \in E - N_1$, there exists at most one $v \in \partial C_f$ with $\rho(v) = y$.

PROPOSITION A4. ρ *is* $\mathscr{C}$*-measurable, and the restriction of* ρ^{-1} *to* $E - N_1$ *is one-to-one and* $\mathscr{B}$*-measurable.*

PROOF. Let

$$h(v, y) = \int_E {}_vP^x\{X_{\zeta-} = y\}\mu_0(dy) = {}_vP^{\mu_0}\{X_{\zeta-} = y\}.$$

By Lemma A3 and Fubini, h is $\mathscr{C} \times \mathscr{B}$-measurable. ${}_vP^x\{X_{\zeta-} = y\}$ is constant on E_v, being one if $y = \rho(v)$ and zero otherwise. By hypothesis (L), either $E_v = \varnothing$ or $\mu_0(E_v) > 0$. If $B \in \mathscr{B}$ and $K \in \mathscr{C}$, $K \subset \partial C_f$, let

$$\begin{aligned} A_{BK} &= \{(v, y): v \in K, y \in B \text{ and } h(v, y) > 0\} \\ &= \{(v, y): v \in K, y \in B \text{ and } y = \rho(v)\}. \end{aligned}$$

Now $\rho^{-1}(B)$ is the projection of $A_{B,\partial C_f}$ on ∂C_f. $A_{B,\partial C_f}$ is a Borel subset of a Polish space and hence is a Lusin space; the projection is a continuous map, and it is one-to-one since each $v \in \partial C_f$ has at most one pole. By a theorem of Lusin [**2**], $\rho^{-1}(B) \in \mathscr{C}$. Thus ρ is measurable. If we restrict ρ^{-1} to $E - N_1$, it becomes one-to-one, and we can project $A_{E-N_1,K}$ on E to conclude that ρ^{-1} is also measurable. Q.E.D.

Thus, for a cap C_f of the above form, define the *Green's function* g of the cap C_f by

$$g(x, y) = v(x), \quad \text{where } v = \rho^{-1}(y),\ x \in E,\ y \in E - N,$$

and

$$N = E - \rho(\rho^{-1}(E - N_1)).$$

THEOREM A5. *$(x, y) \to g(x, y)$ is $\mathscr{B} \times \mathscr{B}$-measurable.*

PROOF. The set $\Gamma_\lambda = \{(x, y) \in E \times N^c: g(x, y) < \lambda\}$ is the projection on $E \times E$ of the $\mathscr{C} \times \mathscr{B} \times \mathscr{B}$-measurable set

$$\{(v, x, y) \in \partial C_f \times E \times N^c: y = \rho(v) \text{ and } v(x) < \lambda\}.$$

For any $(x, y) \in E \times N^c$, there is at most one $v \in \partial C_f$ for which $\rho(v) = y$, so by Lusin's theorem, Γ_λ is $\mathscr{B} \times \mathscr{B}$-measurable. Q.E.D.

With similar reasoning, one can derive the following from A2:

PROPOSITION A6. *Let $\Lambda \in \mathfrak{F}^0$. Then $(x, y) \to {}_{g_y}P^x\{\Lambda\}$ is $\mathscr{B} \times \mathscr{B}$-measurable.*

REMARKS. 1°. The Green's function is just a measurable choice of minimal excessive functions. To emphasize this, we will change notation slightly, and write $g_y(x)$ instead of $g(x, y)$. Thus, for $y \in E - N$, g_y is the unique minimal excessive function v with pole y satisfying $\langle v, f\rangle = 1$. We will make no special effort to exhibit any nice properties of g_y as y varies, for we need only its measurability.

2°. The Green's function depends in general on the cap C_f. However, μ_0 is a probability measure so if $f \equiv 1$ and u is an excessive function with $\|u\| \leqq 1$, then $\langle u, f\rangle \leqq 1$, hence $u \in C_f = C_1$. Thus the cap C_1 is large enough to allow us to represent all bounded excessive functions.

If μ is a measure on $\mathscr{B}$, we denote by $G\mu$ the function $G\mu(x) = \int g_y(x)\mu(dy)$. If $G\mu$ is μ_0-a.e. finite, we call it a *Green's potential.* We will investigate the integral representation of excessive functions, and in particular, of Green's potentials.

THEOREM A7 (MOKOBODZKI). *Let $v \in C_f$. Then there exists a unique subprobability measure ν on $\mathscr{C}$ such that for each $x \in E$*

$$v(x) = \int_{\partial C_f} u(x)\nu(du).$$

This result was not explicitly proved in [15]. What was proved was that there exists a unique subprobability measure ν on ∂C_f such that for every continuous affine function b on C_f,

$$b(v) = \int_{\partial C_f} b(v)\nu(dv).$$

The extension is easy, however. Fix x and write the resolvent R_λ in the form $R_\lambda(x, dy) = r_\lambda(y)\mu_0(dy)$, which we can do by hypothesis (L). Now put $r_{\lambda,n}(y) = r_\lambda(y) \wedge nf(y)$. Then

$$R_\lambda^n u(x) \stackrel{\text{def}}{=} \int u(y) r_{\lambda,n}(y)\mu_0(dy)$$

is affine and continuous on C_f, since the topology of C_f is that induced by $L^1(g\mu_0)$. Thus

$$R_\lambda^n v(x) = \int R_\lambda^n u(x)\nu(du).$$

Let $n \to \infty$; since $f > 0$, $R_\lambda^n \uparrow R_\lambda$, hence by monotone convergence

$$R_\lambda v(x) = \int R_\lambda u(x)\nu(du).$$

Now let $n \to \infty$ and note that ${}_\lambda R_\lambda v(x) \uparrow v(x)$, completing the proof.

The natural question to ask at this point is "when is an excessive function a Green's potential?" The condition given below may seem ad hoc, but it is exactly what we need in many applications.

PROPOSITION A8. *Let $v \in C_f$ and let $K \subset E$ be closed. Then v is a Green's potential with support in K iff*

(A.1) $${}_vP^x\{X_{\zeta-} \in K - N\} = 1, \quad \text{all } x \in E_v.$$

PROOF. The "only if" part is a consequence of Proposition 2.1. Conversely, if v satisfies (A.1) write

$$v = \int_{\partial C_f} u\nu(du).$$

We need only show that ν puts all its mass on $\rho^{-1}(K - N)$, which is exactly $\{g_y, y \in K - N\}$. Letting $A = \rho^{-1}(K - N)$, we have

$$v = v_1 + v_2 = \int_A u\nu(du) + \int_{\partial C_f - A} u\nu(du).$$

Let

$$\Lambda = \{X_{\zeta-} \in N \cup K^c \text{ or } X_{\zeta-} \text{ fails to exist}\}.$$

By Propositions 1.3 and A.2,

$$0 = {}_vP^x\{\Lambda\} = \frac{1}{v(x)} \int_{\partial C_f} u(x)\, {}_uP^x\{\Lambda\}\nu(du).$$

Since ${}_uP^x\{\Lambda\}$ is zero if ν has a pole in $K \cap N$ and is one otherwise, this is just

$$= \frac{1}{v(x)} \int_{\partial C_t - A} u(x)\nu(du) = \frac{v_2(x)}{v(x)}.$$

Thus $v_2 \equiv 0$, and we are done. Q.E.D.

APPENDIX B

The Green's function for the h-transform. Let h be a μ_0-a.e. finite excessive function. h will be fixed in the following. If u is h-excessive, we denote by u^* the excessive function $u^* = R_{E_h} hu$, and if v is excessive, v_* will denote the h-excessive function $v_* = vI_{E_h}/h$. We know the h-transform will satisfy hypothesis (L), since the original process does. We will see here that, in most cases, the h-transform will also satisfy hypothesis (G), and that its Green's function ${}_hg$ derives from the original Green's function by ${}_hg_y = (g_y)_*$.

LEMMA B1. (i) *If u is minimal and h-excessive, u_* is minimal and excessive.*

(ii) *If v is minimal and excessive, and if $v = (v_*)^*$, then v_* is minimal and h-excessive.*

PROOF. In both cases we need only show minimality. If a and b are excessive and $u^* = a + b$, then $u = a_* + b_*$. But this means that a_* and b_* are both proportional to u, since u is minimal. It follows that a and b are proportional to u^*.

Similarly, if c and d are h-excessive and $v_* = c + d$, then, by the additivity of the reduit, $v = R_{E_h}hv_* = R_{E_h}hc + R_{E_h}hd = c^* + d^*$. Thus c^* and d^* are proportional to v; hence c and d are proportional to v_*.

Note that if $h(x) = 0$, $P_t(x, \{h > 0\}) = 0$ for all t; hence the function which is infinite on $\{h > 0\}$ and zero on $\{h = 0\}$ is excessive. It follows that for any excessive function v, $R_{E_h}v$ vanishes on $\{h = 0\}$.

If v is minimal excessive, $\rho(v)$ is the pole of v, if it exists. We use the same notation for h-excessive functions. We denote the reduit operator for h-excessive functions by R^h_A.

LEMMA B2. (i) *If u is h-excessive, minimal and $\rho(u) = y$, then $\rho(u^*) = y$, too.*

(ii) *If v is excessive and minimal and $\rho(v) = y$, and in addition $v = (v_*)^*$, then $\rho(v_*) = y$, too.*

PROOF. Let V be a neighborhood of y. Since $R^h_{V^c}u \neq u$, there exists an h-excessive f such that $f \geqq u$ on V^c and $f(x) < u(x)$ for some $x \in E_h$. Then

$$u^*(x) = R_{E_h}hu(x) = h(x)u(x) \leqq h(x)f(x) = f^*(x).$$

Moreover, $f^* \geqq u^*$ on $V^c \cap E_h$, and therefore on V^c since $v^* = 0$ on E_h^c. Since $R_{V^c}v^* \leqq f \leqq v^*$, it follows that $R_{V^c}v^* \neq v^*$, i.e., $\rho(v^*) = y$.

For part (ii), let V be a neighborhood of y and let $f(x) < v(x)$ for some x. If $x \in E_h$, then we immediately have $f_*(x) < v_*(x)$ and $f_* \geqq v_*$ on V^c. This means that $R^h_{V^c}v_* \neq v_*$, hence $\rho(v_*) = y$. But now, $\{h = \infty\}$ is of μ_0-measure zero, and thus contains no fine open set, and in particular, does not contain $\{f < v\}$. Since $v = 0$ on E_h^c, we conclude $\{f < v\} \cap E_h \neq \varnothing$. Q.E.D.

THEOREM B3. *Let h satisfy*

(B.1) $\quad {}_hP^x\{\zeta < \infty, X_{\zeta-}$ *does not exist or* $X_{\zeta-} \in N\} = 0$, *all* $x \in E_h$.

Then the h-transform satisfies hypothesis (G), *and a version of its Green's function is given by* $\{(g_y)_*, y \in E - N_h\}$, *where* N_h *is an h-left polar Borel set and* $\{g_y, y \in E - N\}$ *is the Green's function for X.*

PROOF. If $y \in E - N$, g_y is the only excessive function with pole at y. Thus, if u and v are h-excessive with $\rho(u) = \rho(v) = y$, u^* and v^* are minimal excessive with pole y by Lemma B2(i). Thus both must be proportional to g_y, and hence to each other. Thus u and v are proportional. It follows that $u = c(g_y)_*$ for some constant c. Furthermore, the set N is h-left-polar, for

$$ {}_hP^x\{S_N < \infty\} = {}_hP^x\{S_N < \zeta\} + {}_hP^x\{S_N = \zeta\} = E^x\{h(X_{S_N})\} = 0, $$

where we have used the facts that $S_N \neq \zeta$ by (B.1) and that N is left polar.

But now, X^h satisfies hypothesis (G), hypothesis (L), and has left limits everywhere. By Theorem A5 it has a Green's function $\{g_y^h, y \in E - N_h'\}$, where N_h' is h-left polar. But N is also h-left polar, and we have seen that if $y \in R - N_h$, where $N_h = N_h' \cup N$, $g_y^h = c(g_y)_*$ for some constant c, so we can replace g_y^h by $\{(g_y)_*, y \in E - N_h\}$.

REFERENCES

1. R. M. Blumenthal and R. K. Getoor, *Markov processes and potential theory*, Academic Press, New York and London, 1968. MR **41** #9348.

2. N. Bourbaki, *Éléments de mathématiques, livre* III. *Topologie générale*, Hermann, Paris. MR **3**, 55.

3. C. Dellacherie, *Capacités et processus stochastiques*, Ergebnisse d. Math. **67**, Springer-Verlag, Berlin, 1972.

4. ———, *Ensembles épais; applications aux processus de Markov*, C. R. Acad. Sci. Paris **266** (1968), A1258–A1261. MR **38** #1745.

5. J. L. Doob, *Conditional Brownian motion and the boundary limits of harmonic functions*, Bull. Soc. Math. France **85** (1957), 431–458. MR **22** #844.

6. ———, *Probability theory and the first boundary value problem*, Illinois J. Math. **2** (1958), 19-36. MR **21** #5242.

7. R. Duncan, *Integral representation of excessive functions*, Pacific J. Math. **39** (1971), 125–144. MR **46** #6482.

8. R.K. Getoor, *Regularity of excessive functions*. II, Ann. Math. Statist. **42** (1971), 2056–2063. MR **45** #9386.

9. R.-M. Hervé, *Recherches axiomatiques sur la théorie des fonctions surharmoniques et du potentiel*, Ann. Inst. Fourier (Grenoble) **12** (1962), 415–571. MR **25** #3186.

10. K. Janssen, *On the existence of a Green function for harmonic spaces*, Math. Ann. **208** (1974), 295-303. MR **50** #2538.

11. P. A. Meyer, *Processus de Markov: La frontière de Martin*, Lecture Notes in Math., No. 77, Springer-Verlag, Berlin and New York, 1968. MR **39** #7669.

12. ———, *Le retournement du temps d'après Chung et Walsh*, Seminaire de Probabilités V (Univ. Strasbourg) Lecture Notes in Math., No. 191, Springer, Berlin, 1971.

13. ———, *Représentation intégrale des fonctions excessives résultats de Mokobodzki*, Séminaire de Probabilités V (Univ. Strasbourg), Lecture Notes in Math., No. 191, Springer, Berlin, 1971. MR **51** #13260.

14. P.-A. Meyer, R. T. Smythe and J. B. Walsh, *Birth and death of Markov processes*, Proc. Sixth Berkeley Sympos. Math. Statist. and Probability, Vol. III, pp. 295–305.

15. G. Mokobodzki, *Dualité formelle et représentation intégrale des fonctions excessives*, Actes du Congress Int. des Mathematiciens 1970, t. 2, 531–535.

16. R. R. Phelps, *Lectures on Choquet's theorem*, Van Nostrand, Princeton, N. J., 1966. MR **33** #1690.

17. C. T. Shih, *On extending potential theory to all strong Markov processes*, Ann. Inst. Fourier (Grenoble) **20** (1970), 303–315. MR **44** #6040.

18. R. T. Smythe and J. B. Walsh, *The existence of dual processes*, Invent. Math. **19** (1973), 113–148. MR **48** #7395.

19. J. C. Taylor, *Duality and the Martin compactification*, Ann. Inst. Fourier (Grenoble) **22** (1972), 95–130. MR **49** #10900.

20. J. B. Walsh and M. Weil, *Representation de temps terminaux et applications aux fonctionnelles additives et aux systèmes de Lévy*, Ann. Sci. Ecole. Norm. Sup. (4) **5** (1972), 121–155. MR **46** #2750.

UNIVERSITY OF BRITISH COLUMBIA

Proceedings of Symposia in Pure Mathematics
Volume 31, 1977

POISSON POINT PROCESS OF BROWNIAN EXCURSIONS AND ITS APPLICATIONS TO DIFFUSION PROCESSES

SHINZO WATANABE

Introduction. The point process of excursions of a diffusion process was studied by Itô [5] in the case when the boundary consists of a single point. In this case, the point process is a *Poisson* point process. Generally, the point process of excursions of a diffusion is not a Poisson point process but a point process of the class [QL] in the sense given in §1. Roughly speaking, a point process is called of the class (QL) if it has a *continuous compensating measure* and, by taking the expectation, we have the so-called *excursion formula*. Such a formula has been obtained, for general Markov processes, by Dynkin [3] and Maisonneuve [7].

Among the excursion point processes of diffusion processes, a particular important role is played by the excursion point process of Brownian motion which we call the *point process of Brownian excursions*. This is a Poisson point process.

In this exposition, we shall study the structure of the Poisson point process of Brownian excursions and, using it, the structure of excursion point process of general diffusion processes. In particular, we will obtain some formulas concerning the stochastic sum over excursions which will be applied to study some class of multiplicative operator functionals of a diffusion solving a system of heat equations with Dirichlet-Neumann boundary conditions. Also, it will be applied to define and describe the *variation of a diffusion along the boundary*.

We will discuss, as another application, the construction of a general class of diffusion processes with boundary conditions. Given a Poisson point process of Brownian excursions, we construct an excursion point process of the diffusion to be constructed and by gluing the excursions using the process on the boundary which is constructed by solving a stochastic differential equation of the jump type based on point process of Brownian excursions, we can define the path functions of the diffusion we want.

AMS (MOS) subject classifications (1970). Primary 60J60, 60J55, 60H05.

1. Point process of the class (QL) and stochastic calculus. Let $(X, \mathscr{B}(X))$ be a measurable space. By a *point function p on X*, we mean a map $p: D_p \subset (0, \infty) \to X$ where the domain D_p is a countable subset of $(0, \infty)$. p defines a counting measure $N_p(dt, dx)$ on $(0, \infty) \times X$ by

$$(1.1) \qquad N_p((0, t] \times U) = \#\{s \in D_p; s \leqq t, p(s) \in U\}, \qquad t > 0, U \in \mathscr{B}(X).$$

A point process is obtained by randomizing the notion of point functions. Let Π_X be the set of all point functions on X and $\mathscr{B}(\Pi_X)$ be the smallest σ-field on Π_X with respect to which all $p \rightsquigarrow N_p((0, t] \times U)$, $t > 0$, $U \in \mathscr{B}(X)$, are measurable. A *point process p* on X is, by definition, a Π_X, $\mathscr{B}(\Pi_X)$-valued random variable; i.e., a measurable map $p: \Omega \ni \omega \rightsquigarrow P(\omega) \in \Pi_X$ defined on a probability space $(\Omega, \mathscr{F}, P)$.

Let $(\Omega, \mathscr{F}, P; \mathscr{F}_t)$ be a probability space with a right-continuous increasing family $\mathscr{F}_t$, $t \in [0, \infty)$, of sub-σ-fields of $\mathscr{F}$. From now on, we consider all the point processes to be defined on the quadruplet $(\Omega, \mathscr{F}, P; \mathscr{F}_t)$. A point process p on X is called *$\mathscr{F}_t$-adapted*, if $\{N_p((0, t] \times U)\}_{U \in \mathscr{B}(X)}$ is $\mathscr{F}_t$-measurable for all $t > 0$.

DEFINITION 1.1. A point process on X is called *of the class* (QL) (with respect to $\mathscr{F}_t$) if

(i) it is $\mathscr{F}_t$-adapted,

(ii) it has a *continuous compensating measure*; to be precise, there exists a σ-finite random measure $\phi_p(dt, dx)$ on $[(0, \infty) \times X, \mathscr{B}(0, \infty) \times \mathscr{B}(X)]$ and a sequence $U_n \in \mathscr{B}(X)$ such that $U_n \nearrow$ and $\bigcup_n U_n = X$, such that, if we set $\Gamma = \{U \in \mathscr{B}(X); \exists\, n, U \subset U_n\}$, then, for every $U \in \Gamma$, $t \rightsquigarrow \phi_p((0, t] \times U)$ is a continuous, $\mathscr{F}_t$-adapted, integrable (i.e., $E(\phi_p((0, t] \times U)) < \infty$) increasing process and $t \rightsquigarrow N_p((0, t] \times U) - \phi_p((0, t] \times U)$ is an $\mathscr{F}_t$-martingale.

Note. ϕ_p is uniquely determined from p.

EXAMPLE 1. Let $(X_t, \mathscr{F}_t)$ be a temporally homogeneous Lévy process (i.e., right continuous process with stationary independent increments) on R^d. Let

$$D_p = \{t \in (0, \infty); X_t \neq X_{t-}\}$$

and, for $t \in D_p$, set $p(t) = X_t - X_{t-} \in R^d \backslash \{0\}$. Then $p(t)$ is a point process on $X = R^d \backslash \{0\}$ of the class (QL) (in fact, stationary Poisson point process by Theorem 1.1) with $\phi_p(dt, dx) = dt \cdot n(dx)$, where $n(dx)$ is the Lévy measure of X_t.

EXAMPLE 2. Let $(X_t, \mathscr{F}_t)$ be a Hunt process on S. Let

$$D_p = \{t \in (0, \infty); X_t \neq X_{t-}\}$$

and, for $t \in D_p$, set $p(t) = (X_{t-}, X_t) \in S \times S \backslash \Delta$ (Δ is the diagonal set). Then $p(t)$ is a point process on $X = S \times S \backslash \Delta$ of the class (QL) with

$$\phi_p((0, t] \times U) = \int_0^t \left[\int_S I_U(X_s, y)\, n(X_s, dy) \right] d\phi_s, \qquad U \in \mathscr{B}(S \times S \backslash \Delta),$$

where $(n(x, dy), \phi_t)$ is a pair of a kernel on $S \times S$ such that $n(x, \{x\}) = 0$, $\forall\, x \in S$, and a continuous increasing additive functional of X_t, called the Lévy system of X_t.

THEOREM 1.1. *Let $n(dx)$ be σ-definite measure on $(X, \mathscr{B}(X))$. If p is a point process on X of the class (QL) such that*

$$\phi_p(dt, dx) = dt \cdot n(dx), \tag{1.2}$$

then for every $t \geqq s \geqq 0$, $\lambda_i > 0$, $U_i \in \mathscr{B}(X)$, $i = 1, 2, \cdots, n$, *such that* $n(U_i) < \infty$ *and are disjoint,*

$$E\left(\exp\left(-\sum_{i=1}^{n}\lambda_i N_p((s, t] \times U_i)\right) \Big/ \mathscr{F}_s\right) = \exp\left\{(t-s)\sum_{i=1}^{n} n(U_i)(e^{-\lambda_i}-1)\right\} \quad a.s. \tag{1.3}$$

Proof is obtained by applying Itô's formula (Theorem 1.4). From this theorem, we see that this p is a stationary Poission point process on X with the characteristic measure n in the sense of Itô [5]. So we call such a p an *$\mathscr{F}_t$-stationary Poisson point process on X with the characteristic measure n.* As is shown in [5], we have

THEOREM 1.2. *Given a σ-finite n on $(X, \mathscr{B}(X))$, we can construct an $\mathscr{F}_t$-stationary Poisson point process on X with the characteristic measure n on a suitable $(\Omega, \mathscr{F}, P; \mathscr{F}_t)$.*

THEOREM 1.3 (STRONG RENEWAL PROPERTY OF POISSON POINT PROCESS). *Let p be an $\mathscr{F}_t$-stationary Poisson point process on X with the characteristic measure n defined on $(\Omega, \mathscr{F}, P; \mathscr{F}_t)$ and let σ be an $\mathscr{F}_t$-stopping time such that $P(\sigma < \infty) = 1$. Let a new point process $\tilde{p}$ be defined by*

$$D_{\tilde{p}} = \{t > 0; t + \sigma \in D_p\} \quad \text{and} \quad \tilde{p}(t) = p(t+\sigma), \qquad t \in D_{\tilde{p}}.$$

Then, if $\tilde{\mathscr{F}}_t = \mathscr{F}_{t+\sigma}$, $\tilde{p}$ is an $\tilde{\mathscr{F}}_t$-stationary Poisson point process with the characteristic measure n. In particular, $\tilde{p}$ is independent of $\tilde{\mathscr{F}}_0 = \mathscr{F}_\sigma$.

This theorem is a direct consequence of Doob's optional sampling theorem.

Now, we discuss some stochastic calculus on point processes, especially on stochastic integrals and Itô's formula. Let p be a point process on X of the class (QL) defined on the space $(\Omega, \mathscr{F}, P; \mathscr{F}_t)$ and $\phi_p(dt, dx)$ be its compensating measure.

DEFINITION 1.2. A real function $f(t, x, \omega)$ defined on $[0, \infty) \times X \times \Omega$ is called *$\mathscr{F}_t$-predictable* if the mapping $(t, x, \omega) \rightsquigarrow f(t, x, \omega)$ is $\mathscr{S}/\mathscr{B}(R)$-measurable, where $\mathscr{S}$ is the smallest σ-field on $[0, \infty) \times X \times \Omega$ with respect to which all g with the following properties are measurable:

(i) for each $t > 0$, $(x, \omega) \rightsquigarrow g(t, x, \omega)$ is $\mathscr{B}(X) \times \mathscr{F}_t$-measurable,

(ii) for each (x, ω), $t \rightsquigarrow g(t, x, \omega)$ is left-continuous.

Let

$$\boldsymbol{F}_p = \Big\{f(t, x, \omega);\ \mathscr{F}_t\text{-predictable and}$$
$$\int_0^{t+}\int_X |f(s, x, \cdot)|\, N_p(ds, dx) < \infty \text{ a.s. } \forall\, t > 0\Big\},$$

$$\boldsymbol{F}_p^1 = \Big\{f(t, x, \omega);\ \mathscr{F}_t\text{-predictable and}$$
$$E\left[\int_0^t\int_X |f(s, x, \cdot)|\, \phi_p(ds, dx)\right] < \infty,\ \forall\, t > 0\Big\},$$

$$\boldsymbol{F}_p^2 = \Big\{f(t, x, \omega);\ \mathscr{F}_t\text{-predictable and}$$
$$E\left[\int_0^t\int_X |f(s, x, \cdot)|^2\, \phi_p(ds, dx)\right] < \infty,\ \forall\, t > 0\Big\},$$

and $F_p^{2,\text{loc}} = \{f(t, x, \omega);\ \mathscr{F}_t$-predictable and $\exists\, T_n$: $\mathscr{F}_t$-stopping times such that $T_n \uparrow \infty$ a.s. and $I_{[0,T_n]}(t) \cdot f(t, x, \omega) \in F_p^2,\ n = 1, 2, \cdots\}$.

For $f \in F_p$, we set

$$(1.4) \qquad P_{f,p}(t) = \int_0^{t+} \int_X f(s, x, \cdot)\, N_p(ds, dx) \qquad \Big(= \sum_{s \leqq t;\, s \in D_p} f(s, p(s), \cdot)\Big).$$

Note that this is an a.s. absolutely convergent sum. For $f \in F_p^1 \cap F_p^2\ (\subset F_p)$, we set

$$(1.5) \qquad \begin{aligned} Q_{f,p}(t) &= P_{f,p}(t) - \int_0^t \int_X f(s, x, \cdot)\, \phi_p(ds, dx) \\ &= \sum_{s \leqq t;\, s \in D_p} f(s, p(s), \cdot) - \int_0^t \int_X f(s, x, \cdot)\, \phi_p(ds, dx). \end{aligned}$$

It is easy to see that $Q_{f,p}(t)$ is a square-integrable, $\mathscr{F}_t$-martingale such that

$$(1.6) \qquad \langle Q_{f,p} \rangle(t) = \int_0^t \int_X f^2(s, x, \cdot)\, \phi_p(ds, dx).$$

This definition is extended to $f \in F_p^2$ by the usual limiting procedure and $Q_{f,p}(t)$ is an $\mathscr{F}_t$-square-integrable martingale. Also, it is extended, further, to $f \in F_p^{2,\text{loc}}$ and $Q_{f,p}(t)$ is then an $\mathscr{F}_t$-locally square-integrable martingale. Sometimes, we denote $Q_{f,p}(t)$ symbolically as

$$(1.7) \qquad Q_{f,p}(t) = \int_0^{t+} \int_X f(s, x, \cdot)\, [N_p(ds, dx) - \phi_p(ds, dx)].$$

THEOREM 1.4 (ITÔ'S FORMULA). *Assume, on $(\Omega, \mathscr{F}, P; \mathscr{F}_t)$, the following is given:*

(i) M_t^i $(i = 1, 2, \cdots, n)$: *$\mathscr{F}_t$-continuous local martingales* $(M_0^i = 0)$.

(ii) ϕ_t^i $(i = 1, 2, \cdots, n)$: *$\mathscr{F}_t$-adapted continuous processes with bounded variation* $(\phi_0^i = 0)$, *i.e., a difference of two $\mathscr{F}_t$-adapted continuous increasing processes.*

(iii) p: *$\mathscr{F}_t$-adapted point process on X of the class (QL) and* $f^i(t, x, \omega) \in F_p$, $g^i(t, x, \omega) \in F_p^{2,\text{loc}}$ $(i = 1, 2, \cdots, n)$ *such that* $f^i(t, x, \omega) g^j(t, x, \omega) \equiv 0\ \forall\, i, j$. *(We have in mind, e.g., the following situation:* $f^i(t, x, \omega) = h^i(t, x, \omega) I_U(x)$ *and* $g^i(t, x, \omega) = h^i(t, x, \omega) I_{U^c}(x)$ *for some* $U \in \mathscr{B}(x)$ *and* $h \in F$.)

(iv) A_0^i $(i = 1, 2, \cdots, n)$: *$\mathscr{F}_0$-measurable random variables.*

Set

$$A_t^i = A_0^i + M_t^i + \phi_t^i + P_{f^i,p}(t) + Q_{g^i,p}(t), \qquad i = 1, 2, \cdots, n,$$

and denote $A_t = (A_t^i)_{i=1}^n$, $f = (f^i)_{i=1}^n$ *and* $g = (g^i)_{i=1}^n$. *Then, for any C^2-class function F on R^n, we have*

$$(1.8) \qquad \begin{aligned} F[A_t] - F[A_0] &= \sum_{i=1}^n \int_0^t F'_{x_i}[A_s]\, dM_s^i + \sum_{i=1}^n \int_0^t F'_{x_i}[A_s]\, d\phi_s^i \\ &\quad + \frac{1}{2} \sum_{i,j=1}^n \int_0^t F''_{x_i x_j}[A_s]\, d\langle M_i, M_j \rangle_s + P_{\xi,p}(t) + Q_{\eta,p}(t) \\ &\quad + \int_0^t \int_X \Big\{ F[A_s + g(s, x, \cdot)] - F[A_s] \\ &\qquad - \sum_{i=1}^n g^i(s, x, \cdot)\, F'_{x_i}[A_s] \Big\}\, \phi_p(ds, dx) \end{aligned}$$

where $\xi(s, x, \omega) = F[A_{s-} + f(s, x, \omega)] - F[A_{s-}]$ *and* $\eta(s, x, \omega) = F[A_{s-} + g(s, x, \omega)] - F[A_{s-}]$, *cf.* **[2]**, **[6]**, **[9]**, **[19]**.

2. Excursion point process of diffusion processes. As a typical example of the point processes of the class (QL) in the theory of Markov processes, we shall discuss the point processes formed of excursions of diffusion processes.

Let $D = \{x = (x_1, x_2, \cdots, x_n); x_n \geqq 0\}$ be the upper half-space of n-dimensional space, $\partial D = \{x \in D; x_n = 0\}$ be its boundary and $\mathring{D} = \{x \in D; x_n > 0\}$ be its interior. We shall consider the following class of diffusion processes on D described by the stochastic differential equation (cf. **[11]**).

$$(2.1)\quad \begin{aligned} dX_t^i &= \sum_{k=1}^{n} \sigma_k^i(X_t)dB_t^k + b^i(X_t)dt + \sum_{l=1}^{n-1} \tau_l^i(X_t)dM_t^l + \beta^i(X_t)d\phi_t, \\ & \qquad\qquad\qquad\qquad\qquad\qquad i = 1, 2, \cdots, n-1, \\ dX_t^n &= \sum_{k=1}^{n} \sigma_k^n(X_t)dB_t^k + b^n(X_t)dt + d\phi_t, \end{aligned}$$

where $\sigma_k^i(x)$, $b^i(x)$, $i, k = 1, 2, \cdots, n$, are functions defined on D and $\tau_l^i(x)$, $\beta^i(x)$, $i, l = 1, 2, \cdots, n-1$, are functions defined on ∂D. Precise formulation is as follows: By a solution of (2.1), we mean a family of adapted processes (X_t, B_t, M_t, ϕ_t) defined on a quadruplet $(\Omega, \mathscr{F}, P; \mathscr{F}_t)$ such that

(i) $t \rightsquigarrow X_t$ is a D-valued continuous process, $t \rightsquigarrow B_t$ is an R^n-valued continuous process, $t \rightsquigarrow M_t$ is an R^{n-1}-valued continuous process and $t \rightsquigarrow \phi_t$ is an $R^+ = [0, \infty)$-valued continuous increasing process such that $B_0 = 0$, $M_0 = 0$, $\phi_0 = 0$ a.s.;

(ii) ϕ_t increases only when $X_t \in \partial D$, i.e.,

$$\int_0^t I_{\partial D}(X_s)\, d\phi_s = \phi_t \text{ a.s.};$$

(iii) (B_t, M_t) is a system of $\mathscr{F}_t$-martingales such that $\langle B^i, B^j\rangle_t = \delta_{ij}t$, $\langle B^i, M^l\rangle = 0$ and $\langle M^l, M^k\rangle_t = \delta_{lk}\cdot\phi_t$, $i, j = 1, 2, \cdots, n$, $l, k = 1, 2, \cdots, n-1$;

(iv) (X_t, B_t, M_t, ϕ_t) satisfies the equation (2.1) where dB_t and dM_t are understood in the sense of martingale stochastic integrals **[6]**.

We set

$$(2.2)\quad a^{ij}(x) = \sum_{k=1}^{n} \sigma_k^i(x)\, \sigma_k^j(x), \qquad i, j = 1, 2, \cdots, n, x \in D,$$

$$(2.3)\quad \alpha^{ij}(x) = \sum_{l=1}^{n-1} \tau_l^i(x)\, \tau_l^j(x), \qquad i, j = 1, 2, \cdots, n-1, x \in \partial D,$$

and assume that

(2.4) all σ, b, τ and β are bounded and Lipschitz continuous and $a^{nn}(x) \geqq c$ for some positive constant c.

It is shown in **[11]** that, for a given Borel probability measure μ on D, there exists a solution of (2.1) such that the law of X_0 coincides with μ and furthermore, the law of any such solution is uniquely determined. The solution defines a diffusion process $X = (X_t)$ on D and it satisfies

$$(2.5)\quad \int_0^t I_{\partial D}(X_s)\, ds = 0 \quad \text{a.s.}$$

Let

$$\sigma_{\partial D} = \inf\{t > 0;\ X_t \in \partial D\} \tag{2.6}$$

and let $X^\circ = (X_{t \wedge \sigma_{\partial D}})$ be the *absorbing barrier diffusion* obtained by stopping X_t on reaching the boundary.

From now on, let $x \in D$ be fixed and consider the solution (X_t, B_t, M_t, ϕ_t) on $(\Omega, \mathscr{F}, P; \mathscr{F}_t)$ such that $X_0 = x$. Let $A(t)$ be the right-continuous inverse of $t \rightsquigarrow \phi_t$;

$$A(t) = \inf\{u;\ \phi_u > t\}. \tag{2.7}$$

Let

$$D_p = \{s \in (0, \infty);\ A(s-) < A(s)\} \tag{2.8}$$

and let

$$\begin{aligned} \mathscr{W}(D) = \{w\colon [0, \infty) \to D, \text{ continuous}, w(0) \in \partial D, \exists\, \sigma(w) > 0 \text{ such that} \\ \text{(i) } t \in (0, \sigma(w)) \Rightarrow w(t) \in \mathring{D}, \text{ (ii) } t \geqq \sigma(w) \Rightarrow w(t) = w(\sigma(w)) \in \partial D\}. \end{aligned} \tag{2.9}$$

$w \in \mathscr{W}(D)$ is called *an excursion on D*. For $s \in D_p$, we define $p(s) \in \mathscr{W}(D)$ by

$$\begin{aligned} [p(s)](t) &= X(t + A(s-)), \qquad 0 \leqq t \leqq A(s) - A(s-), \\ &= X(A(s)), \qquad t \geqq A(s) - A(s-). \end{aligned} \tag{2.10}$$

Thus, we have defined a point process p on $\mathscr{W}(D)$ which is clearly adapted to $\tilde{\mathscr{F}}_t = \mathscr{F}_{A(t)}$. We call *p the excursion point process of the diffusion X*. This is a point process of the class (QL) with respect to $\tilde{\mathscr{F}}_t$ and its compensating measure has the form $Q^{X(A(s))}(dw)\cdot ds$ where $Q^\xi(dw)$, $\xi \in \partial D$, is a system of σ-finite measure on $\mathscr{W}(D)$ which is an X°-Markovian measure corresponding to an entrance law $K^\xi(t, dx) = Q\{w;\ w(t) \in dx\}$ of X°-processes such that $w(0) = \xi$ a.s. $Q^\xi(dw)$. If X is the *reflecting Brownian motion*, i.e., the case

$$\begin{aligned} \sigma^i_k(x) &= \delta^i_k, \qquad b^i = 0,\ i, k = 1, 2, \cdots, n, \\ \tau^i_l(x) &= 0, \qquad \beta^i = 0,\ i, l = 1, 2, \cdots, n-1, \end{aligned} \tag{2.11}$$

the Q^ξ is given explicitly as the unique σ-finite measure on $\mathscr{W}(D)$ such that

$$\begin{aligned} &Q^\xi(w;\ w(t_1) \in E_1, w(t_2) \in E_2, \cdots, w(t_m) \in E_m, \sigma(w) > t_m) \\ &= \int_{E_1} K^\xi(t_1, x_1)\, dx_1 \int_{E_2} p^0(t_2 - t_1, x_1, x_2)\, dx_2 \\ &\cdot \int \cdots \int_{E_m} p^0(t_m - t_{m-1}, x_{m-1}, x_m)\, dx_m, \quad E_i \in \mathscr{B}(\mathring{D}),\ 0 < t_1 < t_2 < \cdots < t_m \end{aligned} \tag{2.12}$$

where

$$K^\xi(t, x) = \prod_{i=1}^{n-1} \frac{1}{\sqrt{2\pi t}} e^{-(x_i - \xi_i)^2/2t} \cdot \frac{\sqrt{2}}{\sqrt{\pi t^3}} x_n e^{-x_n^2/2t}, \qquad t > 0,\ x \in \mathring{D},\ \xi \in \partial D, \tag{2.13}$$

$$p^0(t, x, y) = \prod_{i=1}^{n-1} \frac{1}{\sqrt{2\pi t}} e^{-(x_i - y_i)^2/2t} \cdot \frac{1}{\sqrt{2\pi t}} (e^{-(x_n - y_n)^2/2t} - e^{-(x_n + y_n)^2/2t}), \qquad t > 0,\ x, y \in \mathring{D}. \tag{2.14}$$

In this case, if we define another point process $\tilde{p}$ on $\mathscr{W}(D)$ by

$$
(2.15)\quad \begin{aligned} D_{\tilde{p}} &= D_p \quad \text{and, for } s \in D_{\tilde{p}}, \\ [\tilde{p}(s)](t) &= X(t + A(s-)) - X(A(s-)), && 0 \leqq t \leqq A(s) - A(s-), \\ &= X(A(s)) - X(A(s-)), && t \geqq A(s) - A(s-), \end{aligned}
$$

then the compensator $\phi_{\tilde{p}}(dt, dw)$ has the form $dtQ^0(dw)$ and hence, by Theorem 1.1, it is an $\tilde{\mathscr{F}}_t$-stationary Poisson point process on $\mathscr{W}(D)$ with the characteristic measure Q^0. It is called the *Poisson point process of Brownian excursions.*

Fundamental formulas are as follows (cf. [3], [4], [7], [8]). Let $W = C([0, \infty) \to D)$ be the space of all continuous D-valued functions on $[0, \infty)$ and introduce the following notations;

$$
(2.16)\qquad \rho_t \colon W \to W
$$

defined by $(\rho_t\, w)\,(s) = w(t \wedge s)$ (stopped path),

$$
(2.17)\qquad \theta_t \colon W \to W
$$

defined by $(\theta_t\, w)\,(s) = w(t + s)$ (shifted path),

$$
(2.18)\qquad \rho_{\partial D} \colon W \to W
$$

defined by $(\rho_{\partial D} w)(t) = w(t \wedge \sigma_{\partial D}(w))$ (stopped path on reaching the boundary).

Let X denote the path $t \rightsquigarrow X_t$ which is clearly a W-valued random variable.

Excursion formula I. Let $Z(s)$ be an $\tilde{\mathscr{F}}_t = \mathscr{F}_{A(t)}$-predictable nonnegative process and $f(s, w, w')$ be a nonnegative Borel function on $(0, \infty) \times W \times \mathscr{W}(D)$. Then

$$
(2.19)\quad \begin{aligned} &E\Big\{\sum_{s \leqq t;\, s \in D_p} Z(s) \cdot f(A(s-), \rho_{A(s-)}X, \rho_{\partial D}[\theta_{A(s-)}X])\Big\} \\ &\quad = E\Big\{\int_0^t Z(s)\Big[\int_{\mathscr{W}(D)} f(A(s), \rho_{A(s)}X, w')Q^{X(A(s))}(dw')\Big]\, ds\Big\}. \end{aligned}
$$

Excursion formula II. Let $Z(s)$ be an $\mathscr{F}_t$-well measurable nonnegative process and $f(s, w, w')$ be as above. Then

$$
(2.20)\quad \begin{aligned} &E\Big\{\sum_{s \leqq \phi_t;\, s \in D_p} Z(A(s-)) \cdot f(A(s-), \rho_{A(s-)}X, \rho_{\partial D}[\theta_{A(s-)}X])\Big\} \\ &\quad = E\Big\{\int_0^t Z(s)\Big[\int_{\mathscr{W}(D)} f(s, \rho_s X, w')Q^{X(s)}(dw')\Big]\, d\phi_s\Big\}. \end{aligned}
$$

Let $t > 0$ be fixed and set

$$
(2.21)\quad \begin{aligned} r(t) &= \sup\{s < t;\, X(s) \in \partial D\}, \\ &= 0 \quad \text{if } \{\ \} = \varnothing, \end{aligned}
$$

$r(t)$ is the *last exit time* from ∂D before t. Clearly $r(t) = A(\phi(t) -)$.

Last exit formula. Let $Z(s)$ be an $\mathscr{F}_s$-well measurable nonnegative process and $g(s, s', w, w')$ be a nonnegative Borel function on $(0, \infty) \times (0, \infty) \times W \times \mathscr{W}(D)$. Then

$$
(2.22)\quad \begin{aligned} &E\{Z(r(t)) \cdot g(t - r(t), \rho_{r(t)}X, \rho_{\partial D}[\theta_{r(t)}X])\, I_{\{r(t)>0\}}\} \\ &\quad = E\Big\{\int_0^t Z(s)\Big[\int_{\mathscr{W}(D)} g(t - s, s, \rho_s X, w')\, I_{\{\sigma(w')>t-s\}} Q^{X(s)}(dw')\Big]\, d\phi_s\Big\}. \end{aligned}
$$

(2.22) is obtained from (2.20) by setting $f(s, w, w') = g(t - s, s, w, w')I_{\{(\sigma w')>t-s\}}$ (cf. Maisonneuve [7]).

Now, we shall introduce a notion of *stochastic sum over excursions*. Let $A(t)$ be defined by (2.7). An interval $(A(s-), A(s))$ is called an *excursion interval* if it is nonempty. It is clear that it is nonempty if and only if $s = 0$ and $X_0 \in \mathring{D}$ (and in this case, $A(0-) = 0$ and $A(0+) = \sigma_{\partial D}$) or $s \in D_p$. Let $t > 0$ be fixed and set

$$\boldsymbol{Z}_t = (0, t) \Big\backslash \bigcup_{s \leq \phi(t);\, s \in D_p \cup \{0\}} (A(s-), A(s) \wedge t). \tag{2.23}$$

Then, clearly $\boldsymbol{Z}_t = \{s \in (0, t)\,;\, X_s \in \partial D\}$ and hence, by (2.5), $|\boldsymbol{Z}_t| :=$ Lebesgue measure of $\boldsymbol{Z}_t = 0$ a.s.

Let $\Phi(s) = \Phi(s, \omega)$ be a measurable $\mathscr{F}_t^X$-adapted process where $\mathscr{F}_t^X$ is the completion of $\sigma(X_u; u \leqq t)$. Let $\int_0^t \Phi(s)\, dX^i(s)$ be defined by

$$\begin{aligned} \int_0^t \Phi(s)\, dX^i(s) &= \sum_{k=1}^n \int_0^t \Phi(s)\sigma_k^i(X_s)\, dB_s^k + \int_0^t \Phi(s) b^i(X_s)\, ds \\ &\quad + \sum_{l=1}^{n-1} \int_0^t \Phi(s)\tau_l^i(X_s)\, dM_s^l + \int_0^t \Phi(s)\beta^i(X_s)\, d\phi_s, \end{aligned} \tag{2.24}$$

$$i = 1, 2, \cdots, n,$$

where we set, by convention,

$$\tau_l^n(x) \equiv 0 \quad \text{on } \partial D \qquad \text{and} \qquad \beta^n(x) \equiv 1 \quad \text{on } \partial D. \tag{2.25}$$

Set also

$$\int_0^t \Phi(s)\, d\hat{X}^i = \int_0^t \Phi(s)\, dX^i - \int_0^t \Phi(s) b^i(X_s)\, ds.$$

THEOREM 2.1. *Let $\Phi(u)$ be an $\mathscr{F}_t^X$-adapted measurable process such that $u \rightsquigarrow \Phi(u)$ is right-continuous having left-hand limits and*

$$E\left[\int_0^t \Phi(u)^2\, du\right] + E\left[\int_0^t |\Phi(u)|\, d\phi_u\right] < \infty$$

for every $t > 0$. Then

$$\begin{aligned} &E\Bigg(\sum_{s \in D_p \cup \{0\};\, s \leq \phi(t)} \left(\int_{A(s-)}^{A(s)\wedge t} \Phi(u)\, d\hat{X}^i(u)\right)^2\Bigg) \\ &\quad = E\left(\int_0^t \Phi^2(u) a^{ii}(X(u))\, du\right), \qquad i = 1, 2, \cdots, n. \end{aligned} \tag{2.26}$$

More generally, if $\Psi(u)$ is a similar process,

$$\begin{aligned} &E\Bigg\{\sum_{s \in D_p \cup \{0\};\, s \leq \phi(t)} \left(\int_{A(s-)}^{A(s)\wedge t} \Phi(u)\, d\hat{X}^i(u)\right)\left(\int_{A(s-)}^{A(s)\wedge t} \Psi(u)\, d\hat{X}^j(u)\right)\Bigg\} \\ &\quad = E\left(\int_0^t \Phi(u)\Psi(u) a^{ij}(X(u))\, du\right), \qquad i, j = 1, 2, \cdots, n. \end{aligned}$$

DEFINITION 2.1. Let $f(s) = f(s, \omega)$ be a $\mathscr{B}(0, \infty) \times \mathscr{F}$-measurable function on $(0, \infty) \times \Omega$. We define $\sum^*_{s \leq \phi(t);\, s \in D_p \cup \{0\}} f(s) = I$ if and only if the finite sum $\sum_{s \leq \phi(t);\, A(s) - A(s-) > \varepsilon} f(s)$ converges in probability to I when $\varepsilon \downarrow 0$.

THEOREM 2.2. *Let $\Phi(u)$ be an $\mathscr{F}_t^X$-adapted measurable process such that $u \rightsquigarrow \Phi(u)$ is right-continuous having left-hand limits and*

$$E\left[\int_0^t \Phi(u)^2\, du\right] + E\left[\int_0^t |\Phi(u)|\, d\phi_u\right] < \infty$$

for every $t > 0$. *Then, using the same convention as* (2.25), *we have, for* $i = 1, 2, \cdots, n$,

$$(2.27)\quad \begin{aligned} \sum^*_{s \leqq \phi(t);\, s \in D_p \cup \{0\}} \int_{A(s-)}^{A(s)\wedge t} \Phi(u)\, dX^i(u) &= \int_0^t \Phi(u)\, dX^i(u) \\ &+ \int_0^t \Phi(u) \left[\frac{a^{ni}(X(u))}{a^{nn}(X(u))} - \beta^i(X(u)) \right] d\phi_u \\ &- \sum_{l=1}^{n-1} \int_0^t \Phi(u) \tau_l^i(X(u))\, dM_t^l. \end{aligned}$$

A proof in the case of reflecting Brownian motion (i.e., the case (2.11)) was published in **[15]** and the general case can be reduced to it. Details will be published in **[17]**. As applications, we give the following two examples.

EXAMPLE 2.1 (VARIATION OF THE DIFFUSION ALONG THE BOUNDARY). Consider the diffusion $X = (X_t)$ defined by the equation (2.1). By Itô's formula and Theorem 2.2 we see that, for $f \in C_b^2(D)$,

$$\sum^*_{s \leqq \phi(t);\, s \in D_p \cup \{0\}} [f(X_{t \wedge A(s)}) - f(X_{A(s-)})]$$

exists and

$$V_{\partial D}(f) := f(X_t) - f(X_0) - \sum^*_{s \leqq \phi(t);\, s \in D_p \cup \{0\}} (f(X_{A(s)\wedge t}) - f(X_{A(s-)}))$$

is given by

$$(2.28)\quad \begin{aligned} V_{\partial D}(f) &= \sum_{i=1}^{n-1} \sum_{l=1}^{n-1} \int_0^t \frac{\partial f}{\partial x_i}(X_s) \tau_l^i(X_s)\, dM_s^l \\ &+ \int_0^t \left[\sum_{i=1}^{n-1} \frac{\partial f}{\partial x_i}(X_s) \left\{ \beta^i(X_s) - \frac{a^{ni}(X_s)}{a^{nn}(X_s)} \right\} \right. \\ &\qquad \left. + \frac{1}{2} \sum_{i,j=1}^{n-1} \alpha^{ij}(X_s) \frac{\partial^2 f}{\partial x_i \partial x_j}(X_s) \right] d\phi_s. \end{aligned}$$

It is natural to call $V_{\partial D}(f)$ the *variation of* $t \rightsquigarrow f(X_t)$ *along the boundary*. As a corollary, we see that $V_{\partial D}(f) = 0$ for all $f \in C_b^2(D)$ if and only if $\alpha^{ij}(x) = 0$, $i, j = 1, 2, \cdots, n-1$, and $\beta^i(x) = a^{ni}(x)/a^{nn}(x)$, $i = 1, 2, \cdots, n-1$, identically on ∂D. In this case, it is natural to call the process a *normally reflecting diffusion process*.

EXAMPLE 2.2 (MULTIPLICATIVE OPERATOR FUNCTIONALS). For simplicity, we consider the case of the reflecting Brownian motion, i.e., the case (2.11), though we can discuss a more general case. Then, the equation (2.1) is now

$$(2.29)\quad \begin{aligned} X^i(t) &= X^i(0) + B^i(t), \qquad i = 1, 2, \cdots, n-1, \\ X^n(t) &= X^n(0) + B^n(t) + \phi(t), \end{aligned}$$

and the fomula (2.27) is equivalent to

$$(2.30)\quad \sum^*_{s \leqq \phi(t);\, s \in D_p \cup \{0\}} \int_{A(s-)}^{A(s)\wedge t} \Phi(u)\, dB^i(u) = \int_0^t \Phi(u)\, dB^i(u), \qquad i = 1, 2, \cdots, n-1,$$

$$(2.31)\quad \sum^*_{s \leqq \phi(t);\, s \in D_p \cup \{0\}} \int_{A(s-)}^{A(s)\wedge t} \Phi(u)\, dB^n(u) = \int_0^t \Phi(u)\, dB^n(u) + \int_0^t \Phi(u)\, d\phi(u).$$

Let $A_{ij}^k(x)$, $i, j = 1, 2$, $k = 0, 1, 2, \cdots, n$, be bounded continuous functions on D. Let

$$(2.32)\quad \begin{aligned} r(t) &= \sup\{s \leqq t,\ X(s) \in \partial D\} \\ &= 0, \quad \text{if } \{\ \} = \varnothing. \end{aligned}$$

Then, for fixed t, $r(t) = A(\phi(t)-)$ a.s. and hence, by (2.26),

$$E\left[\left\{\int_{r(t)}^{t} \Phi(u)\, dB^i(u)\right\}^2\right] \leqq E\left[\sum_{s\leqq\phi(t);\, s\in D_p\cup\{0\}} \left\{\int_{A(s-)}^{A(s)\wedge t} \Phi(u)\, dB^i(u)\right\}^2\right] = E\left[\int_0^t \Phi(u)^2\, du\right]. \tag{2.33}$$

We consider the following stochastic differential equation for $M(t) = (M_{ij}(t))_{i,j=1}^2$:

$$\begin{aligned} M_{i1}(t) &= \delta_{i1} + \sum_{k=1}^{n}\sum_{l=1}^{2}\int_0^t M_{il}(s)A_{l1}^k(X(s))\, dB^k(s) \\ &\quad + \sum_{l=1}^{2}\int_0^t M_{il}(s)A_{l1}^0(X(s))\, ds, \\ M_{i2}(t) &= I_{\{\sigma_{\partial D}>t\}}\cdot\left(\delta_{i2} + \sum_{k=1}^{n}\sum_{l=1}^{2}\int_0^t M_{il}(s)A_{l2}^k(X(s))\, dB^k(s)\right. \\ &\quad \left. + \sum_{l=1}^{2}\int_0^t M_{il}(s)A_{l2}^0(X(s))\, ds\right) \\ &\quad + I_{\{\sigma_{\partial D}\leqq t\}}\cdot\left(\sum_{k=1}^{n}\sum_{l=1}^{2}\int_{r(t)}^t M_{il}(s)A_{l2}^k(X(s))\, dB^k(s)\right. \\ &\quad \left. + \sum_{l=1}^{2}\int_{r(t)}^t M_{il}(s)A_{l2}^0(X(s))\, ds\right). \end{aligned} \tag{2.34}$$

By the estimate (2.33), we can easily show that there exists a unique adapted solution $M(t)$ and it is the multiplicative 2×2 matrices in the sense

$$M(t+s) = M(t)\cdot\theta_t[M(s)] \qquad \text{a.s. } t, s > 0, \tag{2.35}$$

where θ_t is the shift operator induced by the process $X(t)$ (cf. **[10]**). By (2.30) and (2.31), if $f_1(x)$ and $f_2(x)$ are in $C_b^2(D)$ such that $\partial f_1/\partial x_n = 0$ and $f_2 = 0$ on ∂D, then

$$\begin{aligned} &\sum_{j=1}^{2} M_{ij}(t)f_j(X(t)) - \sum_{j=1}^{2} M_{ij}(0)f_j(X(0)) \\ &= \left[\sum_{k=1}^{n}\sum_{l,j=1}^{2}\int_0^t M_{il}(s)A_{lj}^k(X(s))f_j(X(s))\, dB^k(s)\right. \\ &\qquad \left. + \sum_{k=1}^{n}\sum_{j=1}^{2}\int_0^t M_{ij}(s)\frac{\partial f_j}{\partial x_k}(X(s))\, dB^k(s)\right] \\ &\quad + \int_0^t \sum_{j=1}^{2} M_{ij}(s)(Lf)_j(X(s))\, ds, \end{aligned} \tag{2.36}$$

where

$$(Lf)_i(x) = \frac{1}{2}\Delta f_i(x) + \sum_{k=1}^{n}\sum_{j=1}^{2} A_{ij}^k(x)\frac{\partial f_j}{\partial x_k}(x) + \sum_{j=1}^{2} A_{ij}^0(x)f_j(x). \tag{2.37}$$

This is a (weak) martingale version of the fact that if $X(0) = x$ and $u(t, x) = E[M(t)f(X(t))]$, then $u(t, x) = (u_1(t, x), u_2(t, x))$ satisfies the following system of heat equations with Dirichlet-Neumann boundary condition:

$$\begin{aligned} \partial u_i/\partial t &= (Lu)_i, & i &= 1, 2, \\ u_i\big|_{t=0} &= f_i, & i &= 1, 2, \\ \partial u_1/\partial x_n\big|_{\partial D} &= 0, & u_2|_{\partial D} &= 0 \end{aligned} \tag{2.38}$$

(cf. H. Airault **[1]**).

3. Construction of diffusion processes with boundary conditions. A class of diffusion processes on D with boundary conditions was constructed in §2 as a solution of the stochastic differential equation (2.1). The diffusion defined by the solution of (2.1) satisfies the following boundary condition of Wentzell:

$$\frac{1}{2}\sum_{i,j=1}^{n-1}\alpha^{ij}(x)\frac{\partial^2 f}{\partial x_i\partial x_j}(x)+\sum_{i=1}^{n-1}\beta^i(x)\frac{\partial f}{\partial x_i}(x)+\frac{\partial f}{\partial x_n}(x)=0 \quad \text{on } \partial D. \tag{3.1}$$

The most general boundary condition of Wentzell is of the following form **[18]**:

$$\begin{aligned}&\frac{1}{2}\sum_{i,j=1}^{n-1}\alpha^{ij}(x)\frac{\partial^2 f}{\partial x_i\partial x_j}(x)+\sum_{i=1}^{n-1}\beta^i(x)\frac{\partial f}{\partial x_i}(x)+\mu(x)\frac{\partial f}{\partial x_n}(x)\\ &\quad+\int_{R\setminus\{0\}}\left[f(x+g(x,u))-f(x)-\sum_{i=1}^{n-1}\frac{\partial f}{\partial x_i}(x)g^i(x,u)\right]\frac{du}{|u|^2}\\ &\quad-\rho(x)(Lf)(x)=0\end{aligned} \tag{3.2}$$

where L is given by

$$Lf(x)=\frac{1}{2}\sum_{i,j=1}^{n-1}a^{ij}(x)\frac{\partial^2 f}{\partial x^i\partial x^j}(x)+\sum_{i=1}^{n}b^i(x)\frac{\partial f}{\partial x^i}(x). \tag{3.3}$$

Thus, (3.1) is the particular case of (3.2) when $\mu(x)=1$, $g(x,u)=0$ and $\rho(x)=0$. The case of general ρ is obtained by a time change (cf. [**11**, II]) but the method of stochastic differential equations cannot cover the general case when μ degenerates on some part of the boundary or $\int I_{\dot{D}}(x+g(x,u))du/|u|^2>0$ for some $x\in\partial D$.

We can given another probabilistic method of constructing a diffusion process with Wentzell's boundary condition by means of the Poisson point process of Brownian excursions, and this method can cover the most general boundary condition of Wentzell. We prepare, on a probability space, a Poisson point process of Brownian excursions and by solving a stochastic differential equation based on the Brownian excursions, we can associate to each of the Brownian excursions an excursion of the diffusion to be constructed. These excursions are glued together to define a sample path of the diffusion and, in doing this, we need the process on the boundary. This can be constructed by solving a stochastic differential equation of jump type based on the Poisson point process of Brownian excursions. For the process thus constructed, the excursion formulas like (2.19) and (2.20) are almost immediately obtained, and using these formulas we can prove the Markovian property of the process. Details will be given in the forthcoming paper **[16]**. (The first draft of the paper was given in **[12]** and **[13]** (cf. also **[14]**).)

References

1. H. Airault, *Resolution stochastique d'un probleme de Dirichlet-Neumann pour des fonctions à valeures vectorielles*, C. R. Acad. Sci. Paris Sér. A-B **280** (1975), A781–A784.

2. C. Doléans-Dade and P.-A. Meyer, *Intégrales stochastiques par rapport aux martingales locales*, Séminaire de Probabilités IV (Univ. Strasbourg, 1968/69), Lecture Notes in Math., vol. 124, Springer-Verlag, Berlin and New York, 1970, pp. 77–107. MR **42** #5313.

3. E. B. Dynkin, *Wanderings of a Markov process*, Teor. Verojatnost. i Primenen. **16** (1971), 409–436 = Theor. Probability Appl. **16** (1971), 401–428. MR **45** #2796.

4. R. K. Getoor and M. J. Sharpe, *Last exit decompositions and distributions*, Indiana Univ.

Math. J. **23** (1973/74), 377–404. MR **48** #12654.

5. K. Itô, *Poisson point processes attached to Markov processes*, Proc. Sixth Berkeley Sympos. on Math. Statist. and Probability, Vol. III (Univ. Calif., 1970/71), Univ. of California Press, Berkeley, 1972, pp. 225–239.

6. H. Kunita and S. Watanabe, *On square integrable martingales*, Nagoya Math. J. **30** (1967), 209–245. MR **36** #945.

7. B. Maisonneuve, *Exit systems*, Ann. Probability **3** (1975), 399–411.

8. B. Maisonneuve and P. -A. Meyer, *Ensembles aléatoires markoviens, homogènes*, Séminaire de Probabilités VIII, Part 2 (Univ. Strasbourg, 1973/74), Lecture Notes in Math., vol. 381, Springer-Verlag, Berlin and New York, 1974, pp. 172–261. MR **50** #1309.

9. P.-A. Meyer, *Un cours sur les integrales stochastiques*, Séminaire de Probabilités X (Univ. Strasbourg 1974/1975), Lecture Notes in Math., vol. 511, Springer-Verlag, Berlin and New York, 1976, pp. 245–400.

10. M. A. Pinsky, *Multiplicative operator functionals and their asymptotic properties*, Advances in Probability and Related Topics (P. Ney and S. Port, editors), Vol. 3, Dekker, New York, 1974, pp. 1–100. MR **51** #4423.

11. S. Watanabe, *On stochastic differential equations for multidimensional diffusion processes with boundary conditions*. I, II, J. Math. Kyoto Univ. **11** (1971), 169–180; 545–551. MR **43** #1291; **44** #4815.

12. ———, *Construction of diffusion processes with Wentzell's boundary conditions by means of Poisson point processes*, Seminar on Probability, Vol. **41** (1975), 23–54. (Japanese)

13. ———, *Construction of diffusion processes with Wentzell's boundary conditions by means of Poisson point processes*, Third USSR-Japan Sympos. on Probability Theory. II, FAN, Tashkent (1975), pp. 311–345.

14. ———, *Construction of diffusion processes by means of Poisson point process of Brownian excursions*, Proc. Third Japan-USSR Sympos. on Probability Theory, Lecture Notes in Math., Springer-Verlag (to appear).

15. ———, *Multiplicative operator functionals of reflecting Brownian motion and stochastic solutions of a system of heat equations*, Seminar on Probability, Vol. **42** (1976), 88–129. (Japanese)

16. ———, *Construction of diffusion processes with Wentzell's boundary conditions by means of Poisson point processes of Brownian excursions*, Proc. Semester on Prob., Banach Center, Warsaw, 1977 (to appear).

17. ———, *Excursion point process of diffusion and stochastic integral*, Proc. Internat. Sympos. on Stochastic Differential Equations (Kyoto) (to appear).

18. A. D. Wentzell, *On boundary conditions for multi-dimensional diffusion processes*, Theor. Probability. Appl. **4** (1959), 164–177.

19. B. Grigelionis, *Stochastic point processes and martingales*, Liet. matem. rink. XV **3** (1975), 101–114. (Russian)

KYOTO UNIVERSITY

Proceedings of Symposia in Pure Mathemtics
Volume 31, 1977

SOME Q-MATRIX PROBLEMS

DAVID WILLIAMS

1. Let I be a countable set. Let $\{P(t)\}$ be a "standard" transition function on I, "honest" in the sense that $P(t)1 = 1, \forall t$. Put $Q \equiv P'(0)$. It is well known that Q exists and that the Doob-Kolmogorov condition

$$\text{(DK)} \qquad 0 \leqq q_{ac} < \infty \qquad (\forall\, a, c: a \neq c)$$

holds. (Recall that a diagonal element q_{ii} of Q may well satisfy $q_{ii} = -\infty$.)

Let E be an *entrance space* for $\{P(t)\}$ in the sense of Doob's famous paper **[2]**. In particular, the following statements hold. The space $E \supseteq I$ and E is a Borel subset of some metric compactification $\bar{E}$ of I. The transition function $\{P(t)\}$ extends to a strong Feller transition function on E so that

$$P(t, \xi, I) = \sum_{j \in I} p_{\xi j}(t) = 1 \qquad (\forall\, \xi \in E, t > 0),$$

$$P(t): B(I) \to C(E) \qquad (\forall\, t > 0).$$

There exists an E-valued, strong Markov process $X = (X_t, \cdots, P^x: x \in E)$, right-continuous on $[0, \infty)$ and with limits from the left, and with transition function $\{P(t)\}$. In general, X will have branch points β at which $P^\beta\{X(0) = \beta\} = 0$. (I have not listed all of the properties of an entrance space. I have merely selected those which we shall need and have tried to put them in widely-known terminology.)

We now further assume that E is what Doob calls a canonical entrance space. This assumption will be clarified at the appropriate stage of the proof of Theorem 1. Doob's paper shows that for many purposes it is best *not* to work with canonical entrance spaces. On this occasion, it is advantageous to do so.

2. Entrance spaces, entrance-exit spaces and the like are not only splendid things in their own right; they are *essential* if we are to understand properly what happens on I. Even so, I regard them as means to an end. If chain theory is to have a life of its own, it has to recover some of the "grittiness" which was once be-

AMS (MOS) subject classifications (1970). Primary 60J25, 60J35, 60J40.

lieved to characterise it. It does not do Doob's marvelous paper justice to regard it as proclaiming the *complete* absorption of chain theory into the glorious Nirvana of "general" theory. A chain is a chain is a chain.

Chains are gritty exactly where they possess local time, that is, at points of I and at regular fictitious states. It is possible to find smoothness properties of excessive functions (for example, results asserting continuity instead of just semicontinuity in important cases) which characterise this grittiness. Once one supplies a tiny missing link in the sequence of excessive function results obtained by Doob **[2]** and Archinard **[1]**, these further smoothness results are straightforward. However, that tiny missing link immediately translates into concrete terms regarding I. I am going to write about these concrete results here. Before doing so, let me emphasise one point: Except for (DK), every necessary condition on Q-matrices mentioned in this paper was first brought to light by local time considerations and only later proved by "topological" arguments.

3. Here is a natural generalisation of the (DK) condition.

THEOREM 1. *Suppose that $A \subseteq I$, $C \subseteq E\backslash A$, A is compact in E and C is closed in E. Then*

$$\sup_{\mu} E^{\mu}J(t, A, C) < \infty \qquad (\forall\, t)$$

where $J(t, A, C) \equiv \#\{s \leqq t: X_{s-} \in A, X_s \in C\}$.

(Here, of course, μ runs over probability measures on E and you know what E^{μ} means.) Purists should note that the assumption $A \subseteq I$ "should" read: For $a \in A$, a is regular (for $\{a\}$).

It is clearly enough to prove

LEMMA 1. *For some $\delta > 0$, $\inf_{a\in A} P^{a}[T_C > \delta] > \delta$, where, as usual, T_C is the hitting time of C. (Then, for every μ,*

$$P^{\mu}[J(\delta, A, C) > n] < (1-\delta)^{n} \quad \textit{and} \quad E^{\mu}J(\delta, A, C) < \delta^{-1}.)$$

IMPORTANT REMARK. Lemma 1 can fail in the case of a general entrance space and even for a canonical *entrance-exit* space. (See the example below.) Excessive function theory shows that the function $a \to P^{a}[T_C > \delta]$ is *upper* semicontinuous and this is the "wrong way round" for us.

PROOF OF LEMMA 1. If the lemma is false, then we can find a sequence (a_n) of points of A such that

$$\lim_{n} P^{a_n}[T_C < \varepsilon] = 1, \qquad \forall\, \varepsilon > 0, \tag{1}$$

and that, in addition, $\lim_n a_n = a_\infty$ exists in A. Because $\{P(t)\}$ is strong Feller,

$$p_{a_n a_\infty}(t) \to p_{a_\infty a_\infty}(t), \qquad \forall\, t > 0,$$

so, by an obvious use of Laplace transforms,

$$\lim_{n} P^{a_n}[T_{a_\infty} < \varepsilon] = 1, \qquad \forall\, \varepsilon > 0. \tag{2}$$

Since

$$P^{a_n}[T_{a_\infty} < T_C < \varepsilon] < P^{a_\infty}[T_C < \varepsilon] \downarrow 0 \qquad (\varepsilon \downarrow 0),$$

we must have from (1) and (2):

$$\lim_n P^{a_n}[T_C < T_{a_\infty} < \varepsilon] = 1, \qquad \forall\, \varepsilon > 0.$$

Hence there exists a sequence (c_n) of points of C such that

$$\lim_n P^{c_n}[T_{a_\infty} < \varepsilon] = 1, \qquad \forall\, \varepsilon > 0,$$

so that

(3) $$\lim_n p_{c_n j}(t) = p_{a_\infty j}(t), \qquad \forall\, j, \forall\, t > 0.$$

Because E is a *canonical* entrance space, (3) implies that $c_n \to a_\infty$. (Indeed, this fact comes near to characterising canonical entrance spaces.) The desired contradiction now follows because C is closed and $a_\infty \notin C$.

EXAMPLE. To see that Lemma 1 can fail for a general entrance space, use the picture

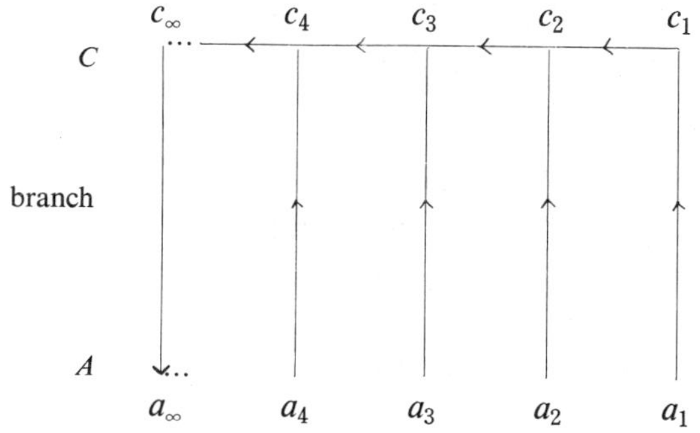

as a guide, make a_∞ absorbing and make c_∞ a branch point with $P^{c_\infty}[X(0) = a_\infty] = 1$. The picture shows the natural *extrance-exit* space for the chain, so the example is in no sense artificial.

4. We need the long-known fact that Q represents the restriction to $I \times \mathscr{B}(I)$ of the Lévy kernel N of X on $E \times \mathscr{B}(E)$. We can then deduce

(LCP) LOCAL CHARACTER PROPERTY **[5]**, **[6]**. *Suppose that C is closed in E and that $a \in I \backslash C$. Then*

$$Q(a, C) \equiv \sum_{c \in C} q_{ac} \leqq N(a, C) < \infty.$$

PROOF. From Theorem 1,

$$\infty > E^a J(t, a, C) = \int_0^t p_{aa}(s) N(a, C)\, ds.$$

On combining (LCP) with the fact that E is Hausdorff, we obtain

(N) $$\sum_{j \notin \{a, b\}} q_{aj} \wedge q_{bj} < \infty \qquad (\forall\, a, b\colon a \neq b)$$

a condition implicit, but nowhere explicit, in Neveu's work on chains.

If X is totally instantaneous, so that

(TI) $$q_{ii} = -\infty \qquad (\forall\, i),$$

then, by right continuity of paths, *no point of I can be isolated in I*. On combining this fact with (LCP), we complete the proof of the "only if" part of the following

THEOREM 2 [5], [6]. *Let Q be an $I \times I$ matrix satisfying* (TI). *Then there exists $\{P(t)\}$ with $P'(0) = Q$ if and only if Q satisfies* (DK), (N) *and also the "safety" condition*

(S) *there exists an infinite set $K \subseteq I$ such that*

$$Q(i, K\backslash i) < \infty, \qquad \forall\, i.$$

Proof of the "if" part of Theorem 1 is much harder and is much more fun. The rather complicated construction used is based on "Kendall's branching procedure", ideas of Feller and Lévy, and a combinatorial lemma due to P. D. Seymour.

In [7], a similar (and more amusing) construction, which owes a lot to Freedman's book [3], is used to settle the existence problem for totally instantaneous chains which satisfy the Kolmogorov backward equations. A lot remains to be done on the *forward equations* and on the important problem of when we can have $N = Q$ (in the obvious sense).

5. Again let $\{P(t)\}$ be a "standard" honest transition function on I and let E be a *canonical* entrance space for $\{P(t)\}$. The topology induced on I is completely determined by $\{P(t)\}$: Indeed $i_n \to i$ if and only if

$$p_{i_n j}(t) \to p_{ij}(t) \qquad (\forall\, j, t).$$

Put $Q \equiv P'(0)$. Note an important consequence of (LCP): *If A is an infinite subset of I and if for some (necessarily unique) a_0 in A,*

$$\inf\{q_{a_0 a}\colon a \in A\backslash a_0\} > 0,$$

then A is compact (with a_0 as its sole point of accumulation). This allows us to obtain "genuinely concrete" applications of the following result.

THEOREM 3. *Suppose that $\{P(t)\}$ is irreducible recurrent. Let m be a (σ-finite) invariant measure for $\{P(t)\}$ so that m is unique up to constant multiples.*

(i) *Suppose that $A \subseteq I$ and A is compact. Then $m(A) < \infty$.*

(ii) *Suppose further that $C \subseteq I\backslash A$ and C is compact. Then*

$$\sum_{a \in A} m_a Q(a, C) < \infty.$$

Theorem 3 is clearly relevant to the problem of obtaining the analogue of Theorem 2 for the situation when (for a given m) $\{P(t)\}$ is restricted to be *m-symmetrisable* in the sense that (see Silverstein [4]) $m_i p_{ij}(t) = m_j p_{ji}(t) \quad (\forall\, i, j, t)$.

SKETCHED PROOF OF THEOREM 3. *Part* (i). Let X^A be the chain obtained by observing X only at those times when it is in A. Then the type of argument used in the proof of Lemma 1 shows that for $t > 0$,

$$P^A(t)\colon B(A) \to C(A), \tag{4}$$

where $\{P^A(t)\}$ is the transition function of X^A. (Thus the compact set A is the unique canonical entrance space for $\{P^A(t)\}$ and X^A is standard in the usual process-theory sense.)

Let (A_n) be a sequence of *finite* subsets of A with $A_n \uparrow A$. From (4) and Dini's

Theorem, for $t > 0$,

$$P^A(t, a, A_n) \uparrow P^A(t, a, A) \qquad (=1)$$

uniformly over $a \in A$. (We could now deduce that each $P^A(t)$ $(t > 0)$ is a *compact* operator on $B(A)$ and so obtain the desired result from ergodic theory. It seems easier to argue as follows.)

Pick n so that $P^A(1, a, A_n) > \frac{1}{2}$ $(\forall\, a \in A)$. By construction of X^A, $m|_A$ is invariant for $\{P^A(t)\}$. Hence

$$\infty > m(A_n) = \sum_{a \in A} m_a P^A(1, a, A_n) > \tfrac{1}{2}\, m(A).$$

Part (ii). Start $X^{A \cup C}$ according to its invariant probability measure $m(A \cup C)^{-1} \cdot m|_{A \cup C}$ and apply Theorem 1.

6. It seems to me that there are good problems galore in this area. If you want an 'abstract' one, how about generalising Theorem 2 from chains to processes? For myself, I prefer the many remaining concrete questions on chains, especially as it is becoming clear that completely new techniques will be needed to tackle some of them.

But you must (and will) decide for yourselves whether anything in this article is worth following up. It is certainly *not* true here that "La distance n'y fait rien; il n'y a que le premier pas qui coûte". This famous quotation comes from a letter from the Marquise du Deffand to d'Alembert. It is a comment on the legend that St. Denis walked two leagues, carrying his head under his arm. Whether or not he could see where he was going is not recorded.

References

1. E. Archinard. *Taboo probabilities in the entrance boundary theory of Markov chains*, Z. Wahrscheinlichkeitstheorie Verw. Gebiete **29** (1974), 165–179. MR **51** #4429.

2. J. L. Doob, *State spaces for Markov chains*, Trans. Amer. Math. Soc. **149** (1970), 279–305. MR **41** #2778.

3. D. A. Freedman, *Approximating countable Markov chains*, Holden-Day, San Francisco, Calif., 1971. MR **45** #1263.

4. M. L. Silverstein, *Symmetric Markov processes*, Lecture Notes in Math., vol. 426, Springer-Verlag, Berlin, 1974.

5. D. Williams, *The Q-matrix problem for Markov chains*, Bull. Amer. Math. Soc. **81** (1975), 1115–1118.

6. ———, *The Q-matrix problem*, Séminaire de Probabilités X (Univ. Strasbourg), Lecture Notes in Math., vol. 511, Springer-Verlag, Berlin and New York, 1976.

7. ———, *The Q-matrix problem*, 2: *The Kolmogorov backward equations*, Séminaire de Probabilités X (Univ. Strasbourg), Lecture Notes in Math., vol. 511, Springer-Verlag, Berlin and New York, 1976.

UNIVERSITY COLLEGE, SWANSEA